Welcome!

Thank you for joining us! As you explore this book, you will find a number of active learning components that help you learn the material at your own pace.

1. **CODE CHALLENGES** ask you to implement the algorithms that you will encounter (in any programming language you like). These code challenges are hosted in the "Bioinformatics Textbook Track" location on Rosalind (http://rosalind.info), a website that will automatically test your implementations.

2. **CHARGING STATIONS** provide additional insights on implementing the algorithms you encounter. However, we suggest trying to solve a Code Challenge before you visit a Charging Station.

3. **EXERCISE BREAKS** offer "just in time" assessments testing your understanding of a topic before moving to the next one.

4. **STOP and Think** questions invite you to slow down and contemplate the current material before continuing to the next topic.

5. **DETOURS** provide extra content that didn't quite fit in the main text.

6. **FINAL CHALLENGES** ask you to apply what you have learned to real experimental datasets.

This textbook powers our popular online courses on Coursera. We encourage you to sign up for a session and learn this material while interacting with thousands of other talented students from around the world. You can also find lecture videos and PowerPoint slides at the textbook website, http://bioinformaticsalgorithms.org.

Bioinformatics Algorithms:
An Active Learning Approach
2nd Edition, Vol. II

Phillip Compeau & Pavel Pevzner

http://bioinformaticsalgorithms.org

©2015

Copyright © 2015 by Phillip Compeau and Pavel Pevzner. All rights reserved.

This book or any portion thereof may not be reproduced or used in any manner whatsoever without the express written permission of the publisher except for the use of brief quotations in a book review.

Printed in the United States of America

Fourth Printing, 2017

ISBN: 978-0-9903746-2-6

Library of Congress Control Number: 2015945208

Active Learning Publishers, LLC
9768 Claiborne Square
La Jolla, CA 92037

To my family. — P. C.

To my parents. — P. P.

In Case You Missed Volume I...

CHAPTER 1

CHAPTER 2

CHAPTER 3

CHAPTER 4

CHAPTER 5

CHAPTER 6

Volume II Overview

CHAPTER 7 — p. 2

CHAPTER 8 — p. 68

CHAPTER 9 — p. 120

CHAPTER 10 — p. 178

CHAPTER 11 — p. 234

Contents

List of Code Challenges   xvii

About the Textbook   xix
    Meet the Authors xix
    Meet the Development Team xx
    Acknowledgments xxi

7 Which Animal Gave Us SARS?   2
    The Fastest Outbreak 3
        Trouble at the Metropole Hotel 3
        The evolution of SARS 3
    Transforming Distance Matrices into Evolutionary Trees 5
        Constructing a distance matrix from coronavirus genomes 5
        Evolutionary trees as graphs 7
        Distance-based phylogeny construction 10
    Toward An Algorithm for Distance-Based Phylogeny Construction 12
        A quest for neighboring leaves 12
        Computing limb lengths 14
    Additive Phylogeny 17
        Trimming the tree 17
        Attaching a limb 19
        An algorithm for distance-based phylogeny construction 19
        Constructing an evolutionary tree of coronaviruses 20
    Using Least Squares to Construct Approximate Distance-Based Phylogenies . 22
    Ultrametric Evolutionary Trees 23
    The Neighbor-Joining Algorithm 27
        Transforming a distance matrix into a neighbor-joining matrix 27

   Analyzing coronaviruses with the neighbor-joining algorithm 31
   Limitations of distance-based approaches to evolutionary tree construction 33
  Character-Based Tree Reconstruction 33
   Character tables 33
   From anatomical to genetic characters 34
   How many times has evolution invented insect wings? 35
  The Small Parsimony Problem 37
  The Large Parsimony Problem 43
  Epilogue: Evolutionary Trees Fight Crime 48
  Detours 51
   When did HIV jump from primates to humans? 51
   Searching for a tree fitting a distance matrix 51
   The four point condition 52
   Did bats give us SARS? 54
   Why does the neighbor-joining algorithm find neighboring leaves? . . . 56
   Computing limb lengths in the neighbor-joining algorithm 61
   Giant panda: bear or raccoon? 62
   Where did humans come from? 62
  Bibliography Notes 66

8 How Did Yeast Become a Wine Maker?   68

  An Evolutionary History of Wine Making 69
   How long have we been addicted to alcohol? 69
   The diauxic shift 70
  Identifying Genes Responsible for the Diauxic Shift 70
   Two evolutionary hypotheses with different fates 70
   Which yeast genes drive the diauxic shift? 71
  Introduction to Clustering 72
   Gene expression analysis 72
   Clustering yeast genes 74
  The Good Clustering Principle 76
  Clustering as an Optimization Problem 78
  Farthest First Traversal 79
  k-Means Clustering 82
   Squared error distortion 82
   k-means clustering and the center of gravity 83
  The Lloyd Algorithm 85

- From centers to clusters and back again 85
- Initializing the Lloyd algorithm 87
- k-means++ Initializer 88
- Clustering Genes Implicated in the Diauxic Shift 89
- Limitations of k-Means Clustering 90
- From Coin Flipping to k-Means Clustering 92
  - Flipping coins with unknown biases 92
  - Where is the computational problem? 95
  - From coin flipping to the Lloyd algorithm 95
  - Return to clustering 96
- Making Soft Decisions in Coin Flipping 97
  - Expectation maximization: the E-step 97
  - Expectation maximization: the M-step 99
  - The expectation maximization algorithm 100
- Soft k-Means Clustering 100
  - Applying expectation maximization to clustering 100
  - Centers to soft clusters 101
  - Soft clusters to centers 102
- Hierarchical Clustering 103
  - Introduction to distance-based clustering 103
  - Inferring clusters from a tree 106
  - Analyzing the diauxic shift with hierarchical clustering ... 108
- Epilogue: Clustering Tumor Samples 109
- Detours 111
  - Whole genome duplication or a series of duplications? 111
  - Measuring gene expression 111
  - Microarrays 112
  - Proof of the Center of Gravity Theorem 113
  - Transforming an expression matrix into a distance/similarity matrix .. 114
  - Clustering and corrupted cliques 115
- Bibliography Notes 118

9 How Do We Locate Disease-Causing Mutations? 120

- What Causes Ohdo Syndrome? 121
- Introduction to Multiple Pattern Matching 122
- Herding Patterns into a Trie 123
  - Constructing a trie 123

Applying the trie to multiple pattern matching	125
Preprocessing the Genome Instead	127
Introduction to suffix tries	127
Using suffix tries for pattern matching	127
Suffix Trees	131
Suffix Arrays	133
Constructing a suffix array	133
Pattern matching with the suffix array	134
The Burrows-Wheeler Transform	136
Genome compression	136
Constructing the Burrows-Wheeler transform	136
From repeats to runs	138
Inverting the Burrows-Wheeler Transform	139
A first attempt at inverting the Burrows-Wheeler transform	139
The First-Last Property	141
Using the First-Last property to invert the Burrows-Wheeler transform	144
Pattern Matching with the Burrows-Wheeler Transform	147
A first attempt at Burrows-Wheeler pattern matching	147
Moving backward through a pattern	148
The Last-to-First mapping	150
Speeding Up Burrows-Wheeler Pattern Matching	153
Substituting the Last-to-First mapping with count arrays	153
Getting rid of the first column of the Burrows-Wheeler matrix	154
Where are the Matched Patterns?	156
Burrows and Wheeler Set Up Checkpoints	157
Epilogue: Mismatch-Tolerant Read Mapping	159
Reducing approximate pattern matching to exact pattern matching	159
BLAST: Comparing a sequence against a database	160
Approximate pattern matching with the Burrows-Wheeler transform	162
Charging Stations	164
Constructing a suffix tree	164
Solving the Longest Shared Substring Problem	167
Partial suffix array construction	169
Detours	170
The reference human genome	170
Rearrangements, insertions, and deletions in human genomes	170
The Aho-Corasick algorithm	170

From suffix trees to suffix arrays	171
From suffix arrays to suffix trees	173
Binary search	176
Bibliography Notes	177

10 Why Have Biologists Still Not Developed an HIV Vaccine? — 178

Classifying the HIV Phenotype	179
How does HIV evade the human immune system?	179
Limitations of sequence alignment	181
Gambling with Yakuza	182
Two Coins up the Dealer's Sleeve	184
Finding CG-Islands	185
Hidden Markov Models	186
From coin flipping to a Hidden Markov Model	186
The HMM diagram	188
Reformulating the Casino Problem	188
The Decoding Problem	191
The Viterbi graph	191
The Viterbi algorithm	194
How fast is the Viterbi algorithm?	195
Finding the Most Likely Outcome of an HMM	196
Profile HMMs for Sequence Alignment	198
How do HMMs relate to sequence alignment?	198
Building a profile HMM	201
Transition and emission probabilities of a profile HMM	203
Classifying proteins with profile HMMs	207
Aligning a protein against a profile HMM	207
The return of pseudocounts	208
The troublesome silent states	209
Are profile HMMs really all that useful?	216
Learning the Parameters of an HMM	217
Estimating HMM parameters when the hidden path is known	217
Viterbi learning	219
Soft Decisions in Parameter Estimation	221
The Soft Decoding Problem	221
The forward-backward algorithm	222
Baum-Welch Learning	225

The Many Faces of HMMs 227
Epilogue: Nature is a Tinkerer and not an Inventor 227
Detours 229
    The Red Queen Effect 229
    Glycosylation 229
    DNA methylation 229
    Conditional probability 230
Bibliography Notes 232

11 Was *T. rex* Just a Big Chicken?   234

Paleontology Meets Computing 235
Which Proteins Are Present in This Sample? 236
Decoding an Ideal Spectrum 237
From Ideal to Real Spectra 241
Peptide Sequencing 244
    Scoring peptides against spectra 244
    Where are the suffix peptides? 246
    Peptide sequencing algorithm 248
Peptide Identification 249
    The Peptide Identification Problem 249
    Identifying peptides in the unknown *T. rex* proteome 250
    Searching for peptide-spectrum matches 251
Peptide Identification and the Infinite Monkey Theorem 252
    False discovery rate 252
    The monkey and the typewriter 254
    Statistical significance of a peptide-spectrum match 255
Spectral Dictionaries 258
T. rex Peptides: Contaminants or Treasure Trove of Ancient Proteins? 261
    The hemoglobin riddle 261
    The dinosaur DNA controversy 264
Epilogue: From Unmodified to Modified Peptides 264
    Post-translational modifications 264
    Searching for modifications as an alignment problem 265
    Building a Manhattan grid for spectral alignment 267
    Spectral alignment algorithm 271
Detours 273
    Gene prediction 273

　　　　Finding all paths in a graph 274
　　　　The Anti-Symmetric Path Problem 275
　　　　Transforming spectra into spectral vectors 276
　　　　The infinite monkey theorem 278
　　　　The probabilistic space of peptides in a spectral dictionary 278
　　　　Are terrestrial dinosaurs really the ancestors of birds? 279
　　　　Solving the Most Likely Peptide Vector Problem 280
　　　　Selecting parameters for transforming spectra into spectral vectors . . . 281
　　Bibliography Notes 283

Bibliography   **285**

Image Courtesies   **291**

List of Code Challenges

Chapter 7   **2**
  (7A) Compute Distances Between Leaves 11
  (7B) Compute Limb Lengths in a Tree 17
  (7C) Implement **ADDITIVEPHYLOGENY** 20
  (7D) Implement **UPGMA** 25
  (7E) Implement **NEIGHBORJOINING** 30
  (7F) Implement **SMALLPARSIMONY** 40
  (7G) Adapt **SMALLPARSIMONY** to Unrooted Trees 42
  (7H) Find the Nearest Neighbors of a Tree 45
  (7I) Implement **NEARESTNEIGHBORINTERCHANGE** 47

Chapter 8   **68**
  (8A) Implement **FARTHESTFIRSTTRAVERSAL** 80
  (8B) Compute the Squared Error Distortion 82
  (8C) Implement the Lloyd Algorithm for k-Means Clustering 85
  (8D) Implement the Soft k-Means Clustering Algorithm 103
  (8E) Implement **HIERARCHICALCLUSTERING** 106

Chapter 9   **120**
  (9A) Construct a Trie from a Collection of Patterns 124
  (9B) Implement **TRIEMATCHING** 126
  (9C) Construct the Suffix Tree of a String 132
  (9D) Find the Longest Repeat in a String 132
  (9E) Find the Longest Substring Shared by Two Strings 133
  (9F) Find the Shortest Non-Shared Substring of Two Strings 133
  (9G) Construct the Suffix Array of a String 133
  (9H) Implement **PATTERNMATCHINGWITHSUFFIXARRAY** 135

(9I) Construct the Burrows-Wheeler Transform of a String 138
(9J) Reconstruct a String from its Burrows-Wheeler Transform 147
(9K) Generate the Last-to-First Mapping of a String 151
(9L) Implement **BWMatching** 151
(9M) Implement **BetterBWMatching** 156
(9N) Find All Occurrences of a Collection of Patterns in a String 158
(9O) Find All Approximate Occurrences of a Collection of Patterns in a String 162
(9P) Implement **TreeColoring** 169
(9Q) Construct the Partial Suffix Array of a String 169
(9R) Construct a Suffix Tree from a Suffix Array 172

Chapter 10   178
(10A) Compute the Probability of a Hidden Path 190
(10B) Compute the Probability of an Outcome Given a Hidden Path 191
(10C) Implement the Viterbi Algorithm 195
(10D) Compute the Probability of a String Emitted by an HMM 197
(10E) Construct a Profile HMM 206
(10F) Construct a Profile HMM with Pseudocounts 209
(10G) Perform a Multiple Sequence Alignment with a Profile HMM 212
(10H) Estimate the Parameters of an HMM 219
(10I) Implement Viterbi Learning 220
(10J) Solve the Soft Decoding Problem 223
(10K) Implement Baum-Welch Learning 226

Chapter 11   234
(11A) Construct the Graph of a Spectrum 239
(11B) Implement **DecodingIdealSpectrum** 240
(11C) Convert a Peptide into a Peptide Vector 245
(11D) Convert a Peptide Vector into a Peptide 245
(11E) Sequence a Peptide 249
(11F) Find a Highest-Scoring Peptide in a Proteome against a Spectrum . . . 250
(11G) Implement **PSMSearch** 252
(11H) Compute the Size of a Spectral Dictionary 258
(11I) Compute the Probability of a Spectral Dictionary 260
(11J) Find a Highest-Scoring Modified Peptide against a Spectrum 273

About the Textbook

Meet the Authors

PHILLIP COMPEAU is an Assistant Teaching Professor in the Computational Biology Department at Carnegie Mellon University. He is a former postdoctoral researcher in the Department of Computer Science & Engineering at the University of California, San Diego, where he received a Ph. D. in mathematics. He is passionate about the future of both offline and online education, having cofounded Rosalind with Nikolay Vyahhi in 2012. A retired tennis player, he dreams of one day going pro in golf.

PAVEL PEVZNER is Ronald R. Taylor Professor of Computer Science at the University of California, San Diego. He holds a Ph. D. from Moscow Institute of Physics and Technology, Russia and an Honorary Degree from Simon Fraser University. He is a Howard Hughes Medical Institute Professor (2006), an Association for Computing Machinery Fellow (2010), and an International Society for Computational Biology Fellow (2012). He has authored the textbooks *Computational Molecular Biology: An Algorithmic Approach* (2000) and *An Introduction to Bioinformatics Algorithms* (2004) (jointly with Neil Jones).

Meet the Development Team

VU NGO is a Ph. D student in Bioinformatics and Systems Biology at UCSD. He holds a B. S in Biotechnology from Rutgers University. His research interests include epigenetics, gene regulatory networks, and machine learning algorithms. Outside of research, Vu enjoys movies, music and badminton.

MAX SHEN is a student in the Computational and Systems Biology Ph. D. program at the Massachusetts Institute of Technology, having received a B. S. from UCSD. He is excited about the ways algorithms and machine learning are applied in bioinformatics. Outside of research, Max enjoys hip hop dance and playing video games competitively.

JEFFREY YUAN is a graduate student in the Bioinformatics and Systems Biology Program at UCSD. He holds a Sc. B. in Computational Biology from Brown University. His research interests include genome assembly with long reads, the 3-dimensional organization of genome structure, and genomic variation. Outside of research, Jeffrey enjoys reading, board games, and volleyball.

Acknowledgments

This textbook was greatly improved by the efforts of a large number of individuals, to whom we owe a debt of gratitude.

The development team (Vu Ngo, Max Shen, and Jeffrey Yuan), as well as Ksenia Krasheninnikova, implemented coding challenges and exercises, rendered figures, helped typeset the text, and offered insightful feedback on the manuscript.

Glenn Tesler provided thorough chapter reviews and caught many errors in our manuscript.

Sangtae Kim, Seungjin Na, Mihai Pop, and Héctor Corrada Bravo provided many thoughtful comments. Sangtae and Seungjin also generated some images used in the book.

Randall Christopher brought to life our ideas for illustrations in addition to the textbook cover.

Nikolay Vyahhi led a team composed of Andrey Balandin, Artem Suschev, Aleksey Kladov, and Kirill Shikhanov, who worked hard to support an online, interactive version of this textbook used in our online course on Coursera.

Laurence Bernstein and Kai Zhang worked to implement many of the problems in Chapter 9.

Our students on Coursera, especially Mark Mammel and Erika Ramírez, found hundreds of typos in our preliminary manuscript.

Howard Hughes Medical Institute, the Russian Ministry of Education and Science, and the National Institutes of Health generously gave their support for the development of the online course based on this textbook. The Bioinformatics and Systems Biology Program and the Computer Science & Engineering Department at the University of California, San Diego provided additional support.

Finally, our families gracefully endured the many long days and nights that we spent poring over manuscripts, and they helped us preserve our sanity along the way.

P. C. and P. P.
San Diego
July 2015

The Fastest Outbreak

Trouble at the Metropole Hotel

On February 21, 2003, a Chinese doctor named Liu Jianlun flew to Hong Kong to attend a wedding and checked into Room 911 of the Metropole Hotel. The next day, he became too ill to attend the wedding and was admitted to a hospital. Two weeks later, Dr. Jianlun was dead.

On his deathbed, Jianlun told doctors that he had recently treated sick patients in Guangdong Province, China where a deadly, highly contagious respiratory illness had infected hundreds of people. The Chinese government had made brief mention of this incident to the World Health Organization but had concluded that the likely culprit was a common bacterial infection.

By the time anyone realized the severity of the disease, it was already too late to stop the outbreak. On February 23, a man who had stayed across the hall from Dr. Jianlun at the Metropole traveled to Hanoi and died after infecting 80 people. On February 26, a woman checked out of the Metropole, traveled back to Toronto, and died after initiating an outbreak there. On March 1, a third guest was admitted to a hospital in Singapore, where sixteen additional cases of the illness arose within two weeks.

Consider that it took four years for the Black Death, which killed over a third of all Europeans in the 14th Century, to travel from Constantinople to Kiev. Or that HIV took two decades to circle the globe. In contrast, this mysterious new disease had crossed the Pacific Ocean within a week of entering Hong Kong.

As health officials braced for the impact of the fastest-traveling pandemic in human history, panic set in. Businesses were closed, sick passengers were removed from airplanes, and Chinese officials threatened to execute infected patients who violated quarantine.

International travel may have helped the disease spread rapidly, but international collaboration would eventually contain it. In a matter of a few weeks, biologists identified a virus that had caused the epidemic and sequenced its genome. In the process, the mysterious new disease earned a name: **Severe Acute Respiratory Syndrome**, or **SARS**.

The evolution of SARS

The virus causing SARS belongs to a family of viruses called **coronaviruses**, which are named after the Latin *corona* (meaning "crown") because the virus particle resembles the sun's corona (Figure 7.1). Coronaviruses infect the respiratory tracts of mammals and birds but typically cause only minor problems, like the common cold. Before SARS, no one believed that a coronavirus could wreak such havoc.

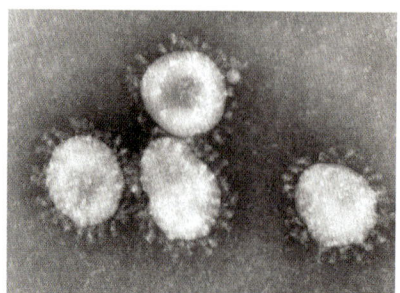 

Figure 7.1 (Left) Coronavirus particles. (Right) A solar eclipse with the sun's corona visible.

Coronaviruses, influenza viruses, and HIV are all **RNA viruses**, meaning that they possess RNA instead of DNA. RNA replication has a higher error rate than DNA replication, and so RNA viruses are capable of mutating more quickly into divergent strains. The rapid mutation of RNA viruses explains why the flu shot changes from year to year and why there are many different subtypes of HIV.

SARS researchers initially hypothesized that, like HIV and influenza, the **SARS coronavirus** (abbreviated as **SARS-CoV**) had jumped from animals to humans. They first named birds as the likely suspect because of the similarities between SARS and "bird flu", a form of influenza originating in chickens that is difficult to transmit to humans but is even deadlier than SARS, killing over half of the people it infects. Yet when researchers sequenced the 29,751 nucleotide-long SARS-CoV genome in April 2003, it became evident that SARS did not come from birds because its genome did not resemble avian coronaviruses.

By fall 2003, researchers had sequenced many SARS-CoV strains from patients in various countries, but many questions still remained unanswered. How did SARS-CoV cross the species barrier to humans? When and where did it happen? How did SARS spread around the world, and who infected whom?

Each of these questions about SARS is ultimately related to the problem of constructing **evolutionary trees** (also known as **phylogenies**). For another example, by constructing an evolutionary tree of primate viruses related to HIV (Figure 7.2), scientists inferred that HIV was transmitted to humans on five separate occasions (see **PAGE 51** **DETOUR: When Did HIV Jump From Primates to Humans?**). But what algorithm did they use to construct this phylogeny?

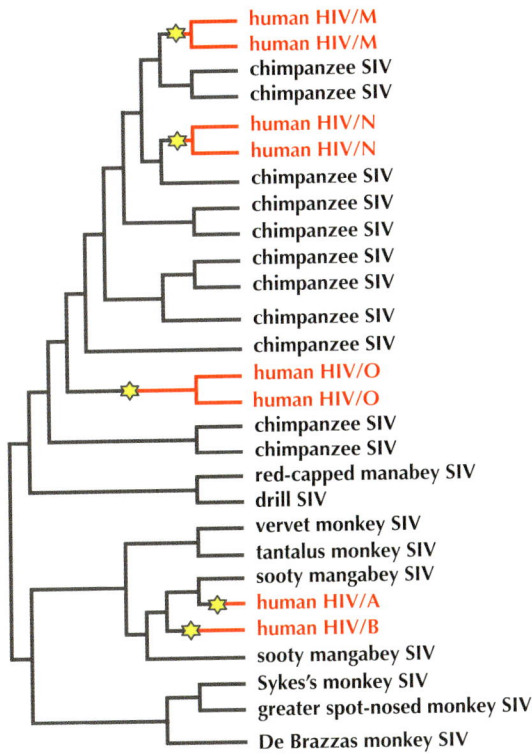

FIGURE 7.2 HIV comprises five different viral families, denoted as A, B, M, N, and O, with the M family responsible for 95% of all HIV infections. The five families are different offshoots of the evolutionary tree for Simian Immunodeficiency Virus (SIV), which infects primates. Stars indicate viruses transitioning from primates to humans. The A and B families originated in sooty mangabey monkeys, whereas the M, N, and O families originated in chimpanzees.

Transforming Distance Matrices into Evolutionary Trees

Constructing a distance matrix from coronavirus genomes

To determine how SARS jumped from animals to humans, scientists started sequencing coronaviruses from various species in order to determine which one is the most similar to SARS-CoV. However, constructing a multiple alignment of entire viral genomes is tricky because viral genes are often rearranged, inserted, and deleted. For this reason, scientists focused on only one of the six genes in SARS-CoV. This gene encodes the **Spike protein**, which identifies and binds to receptor sites on the host's cell membrane.

In SARS-CoV, the Spike protein is 1,255 amino acids long and has rather weak similarity with Spike proteins in other coronaviruses. However, even these subtle similarities turned out to be sufficient for constructing a multiple alignment of Spike proteins across various coronaviruses.

After constructing a multiple alignment of genes from n different species, biologists often transform this alignment into an $n \times n$ **distance matrix** D. In many cases, $D_{i,j}$ represents the number of differing symbols between the genes representing rows i and j of the alignment (Figure 7.3). However, distance matrices can be constructed using a variety of different distance functions in order to suit different applications. For example, $D_{i,j}$ could represent the edit distance between genes from the i-th and j-th species. Or, a distance matrix for n genomes could be constructed from the 2-break distances between each pair of genomes.

Regardless of which distance function we use, in order to be a distance matrix, D must satisfy three properties. It must be **symmetric** (for all i and j, $D_{i,j} = D_{j,i}$), **non-negative** (for all i and j, $D_{i,j} \geq 0$) and satisfy the **triangle inequality** (for all i, j, and k, $D_{i,j} + D_{j,k} \leq D_{i,k}$).

EXERCISE BREAK: Prove that if $D_{i,j}$ is equal to the number of differing symbols between rows i and j of a multiple alignment, then D is symmetric, non-negative, and satisfies the triangle inequality.

Species	Alignment	Distance Matrix			
		Chimp	Human	Seal	Whale
Chimp	ACGTAGGCCT	0	3	6	4
Human	ATGTAAGACT	3	0	7	5
Seal	TCGAGAGCAC	6	7	0	2
Whale	TCGAAAGCAT	4	5	2	0

FIGURE 7.3 A multiple alignment of hypothetical DNA sequences from four species, along with the distance matrix produced by counting the number of differing symbols between each pair of rows in this multiple alignment.

WHICH ANIMAL GAVE US SARS?

By the end of 2003, bioinformaticians had sequenced many coronaviruses taken from a variety of animals and SARS patients and then computed the associated distance matrix. They needed to use this information in order to construct a coronavirus phylogeny and understand the origin and spread of the SARS epidemic.

Evolutionary trees as graphs

You may have noticed that the HIV tree in Figure 7.2 has the structure of a graph. Furthermore, Figure 7.4 (top) shows the representation of the phylogeny of all life as a graph.

Graphs that are used to model phylogenies share two properties. They are connected (i.e. it is possible to reach any node from any other node), and they contain no cycles. For this reason, we will define a **tree** as a connected graph without cycles (see Figure 7.4 (bottom) for a few additional examples).

Take another look at Figure 7.4 (top). You will see that present-day species have been assigned to the **leaves** of the tree, or nodes having degree 1 (in Chapter 3, we defined the degree of a node as the number of edges connected to that node). Nodes with degree larger than 1 are called **internal nodes** and represent unknown ancestor species. Given a leaf j, there is only one node connected to j by an edge, which we call the **parent** of j, denoted PARENT(j). An edge connecting a leaf to its parent is called a **limb**.

> **EXERCISE BREAK:** Prove the following statements:
>
> - Every tree with at least two nodes has at least two leaves.
>
> - Every tree with n nodes has $n-1$ edges.

In a **rooted tree**, one node is designated as a special node called the **root**, and the edges in the tree automatically inherit an implicit orientation away from the root, which is placed at the top or left of the tree (Figure 7.5). This edge orientation models time: the ancestor of all species in the tree is found at the root, and evolution proceeds from the root outward through the tree. Trees without a designated root are called **unrooted**.

> **STOP and Think:** Where would you place the root in the phylogeny in Figure 7.2?

CHAPTER 7

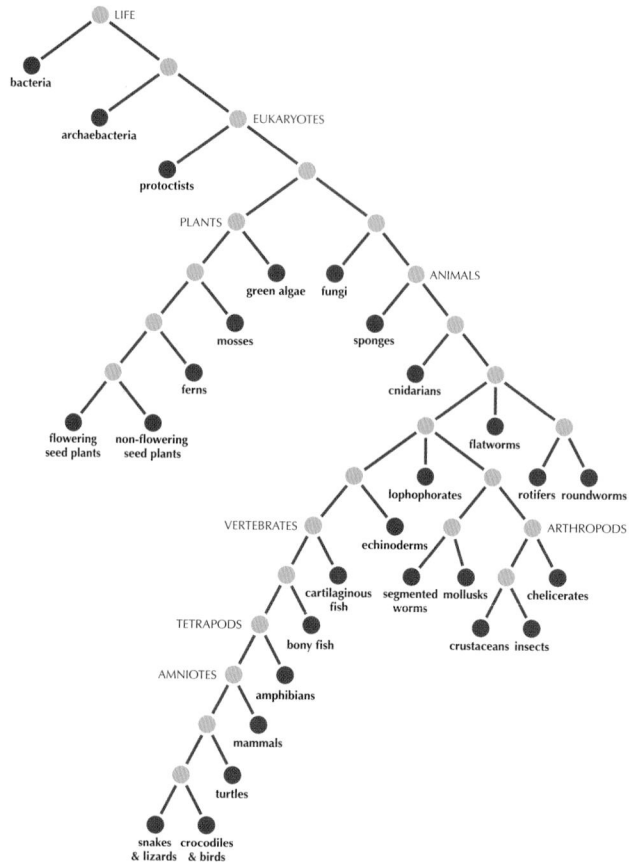

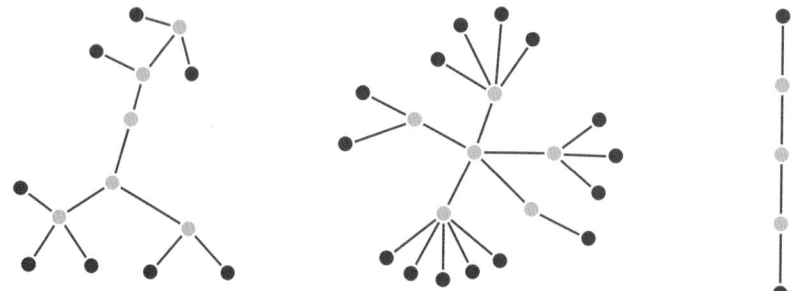

FIGURE 7.4 (Top) A connected acyclic graph that models an evolutionary tree of life on Earth. Present-day species are shown as darker nodes (leaves). (Bottom) Trees come in a variety of different shapes. In each of the three trees shown, leaves (i.e., nodes of degree 1) have been drawn darker than internal nodes (i.e., nodes of larger degree).

We will analyze rooted trees when we attempt to infer the node corresponding to the ancestor of all species in the tree; otherwise, we will analyze unrooted trees. Figure 7.6 shows an unrooted tree of HIV viruses produced from a different dataset than the one used to create Figure 7.2. By proposing two additional subtypes of HIV, it illustrates that the classification of HIV into five families shown in Figure 7.2 is not written in stone.

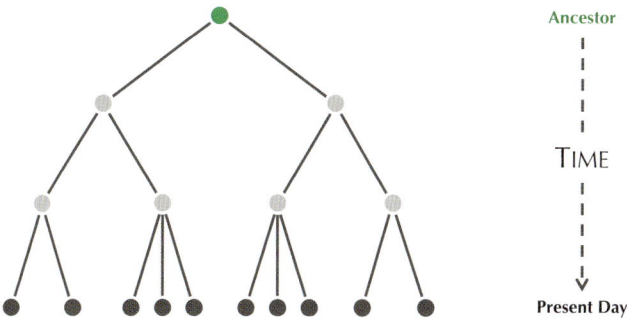

FIGURE 7.5 A rooted tree, with the root (representing an ancestor of all species in the tree) indicated in green at the top of the tree. The presence of the root implies an orientation of edges in the tree away from the root.

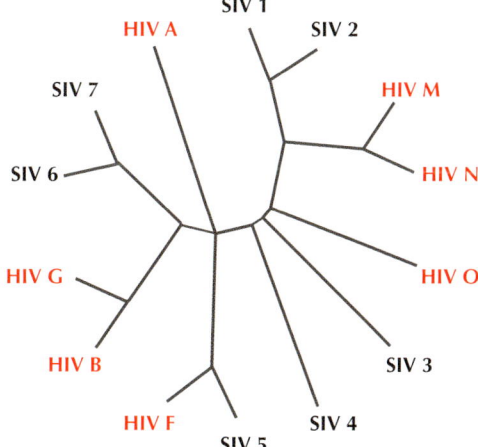

FIGURE 7.6 An unrooted tree of HIV and SIV viruses that suggests additional viral families F and G in addition to the viral families A, B, M, N, and O shown in Figure 7.2.

CHAPTER 7

Distance-based phylogeny construction

We will first focus on deriving an unrooted tree from a distance matrix. The leaves of this tree should correspond to the species represented by the matrix (with internal nodes corresponding to unknown ancestral species). To reflect the evolutionary distance between species in a tree, we assign each edge a non-negative length representing the distance between the organisms that the edge connects, as shown in Figure 7.7.

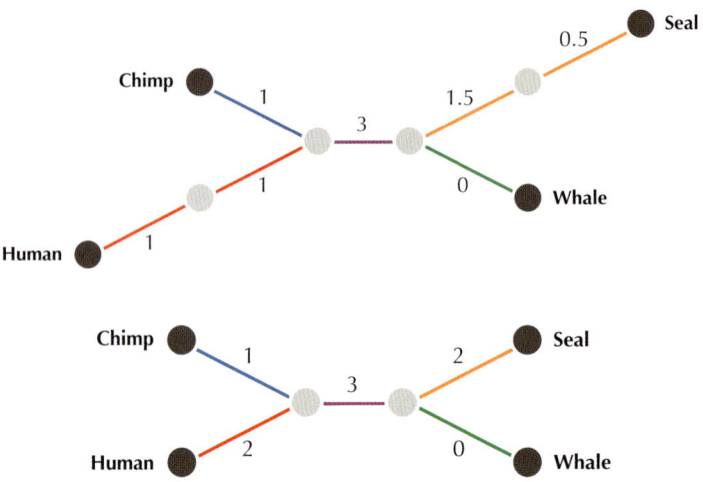

FIGURE 7.7 Two unrooted trees fitting the distance matrix in Figure 7.3. Each of the five maximal non-branching paths in the tree on the top is shown using a different color. Replacing each maximal non-branching path in this tree with a single edge (of length equal to the total length of edges) results in the simple tree shown on the bottom.

> **EXERCISE BREAK:** Prove that there exists exactly one path connecting every pair of nodes in a tree. Hint: what would happen if there were two different paths connecting a pair of nodes? What would happen if there were no paths connecting a pair of nodes?

In this chapter, we define the length of a path in a tree as the sum of the lengths of its edges (rather than the number of edges on the path). As a result, the evolutionary distance between two present-day species corresponding to leaves i and j in a tree T is equal to the length of the unique path connecting i and j, denoted $d_{i,j}(T)$.

WHICH ANIMAL GAVE US SARS?

Distances Between Leaves Problem:

Compute the distances between leaves in a weighted tree.

    Input: A weighted tree with n leaves.
    Output: An $n \times n$ matrix $(d_{i,j})$, where $d_{i,j}$ is the length of the path between leaves i and j.

The Distance Between Leaves Problem is straightforward to solve, but we would like to solve the reverse problem, in which we must construct an unrooted tree that models a given distance matrix. We say that a weighted unrooted tree T **fits** a distance matrix D if $d_{i,j}(T) = D_{i,j}$ for every pair of leaves i and j.

Distance-Based Phylogeny Problem:

Reconstruct an evolutionary tree fitting a distance matrix.

    Input: A distance matrix.
    Output: A tree fitting this distance matrix.

STOP and Think: Does the Distance-Based Phylogeny Problem always have a solution?

Not every distance matrix has a tree fitting it (see **DETOUR: Searching for a Tree Fitting a Distance Matrix**). We therefore call a distance matrix **additive** if there exists a tree that fits this matrix and **non-additive** otherwise. The term "additive" is used because the lengths of all edges along the path between leaves i and j in a tree fitting the matrix D add to $D_{i,j}$.

Note that both trees in Figure 7.7 fit the distance matrix from Figure 7.3, so it would be nice to have a notion of a "canonical" tree fitting a distance matrix. Extending definitions introduced in Chapter 3 to undirected graphs, we say that a path in a tree is **non-branching** if every node other than the beginning and ending node of the path has degree equal to 2. A non-branching path is **maximal** if it is not a subpath of an even longer non-branching path. If we substitute every maximal non-branching path by a single edge whose length is equal to the length of the path, then the tree in Figure 7.7 (top) becomes the tree in Figure 7.7 (bottom). In general, after such a transformation, there are no nodes of degree 2; a tree satisfying this property is called a **simple tree**. It

CHAPTER 7

turns out that if a matrix is additive, then there exists a *unique* simple tree fitting this matrix. In the Distance-Based Phylogeny Problem, we will therefore use the terminology TREE(D) to denote the simple tree fitting the additive distance matrix D. Our question, then, is how to construct TREE(D) from D.

> **EXERCISE BREAK:** Prove that every simple tree with n leaves has at most $n - 2$ internal nodes.

Toward An Algorithm for Distance-Based Phylogeny Construction

A quest for neighboring leaves

A natural first step for solving the Distance-Based Phylogeny Problem would be to ensure that the two closest species with respect to the distance matrix D correspond to **neighbors** in TREE(D). In other words, the minimum value $D_{i,j}$ should correspond to leaves i and j having the same parent. In the rest of this chapter, when we refer to the minimum element of a matrix, we are referring to a minimum **off-diagonal** element, i.e., a value $D_{i,j}$ such that $i \neq j$.

Theorem. *Every simple tree with at least three nodes has a pair of neighboring leaves.*

Proof. Given a simple tree T with at least three nodes, consider a path $P = (v_1, \ldots, v_k)$ that has the maximum number of nodes of any path in T. Because T has at least three nodes, k must be at least 3. Furthermore, nodes v_1 and v_k must be leaves, since otherwise we could extend P into a longer path. Because T is simple, node v_2, which is the parent of v_1, must have at least three adjacent nodes: v_1, v_3, and yet another node w.

We claim that w is a leaf, which would imply that leaves v_1 and w are neighbors. We will proceed by contradiction: if w were not a leaf, then since T is simple, w would be adjacent to another node u. As a result, we could form the path $P' = (u, w, v_2, v_3, \ldots, v_k)$, which contains $k + 1$ nodes and contradicts our original assumption that P has the maximum number of nodes. Thus, w must be a leaf, implying that v_1 and w are neighbors. □

Figure 7.8 (top) illustrates that for neighboring leaves i and j sharing a parent node m, the following equality holds for every other leaf k in the tree:

$$d_{k,m} = \frac{(d_{i,m} + d_{k,m}) + (d_{j,m} + d_{k,m}) - (d_{i,m} + d_{j,m})}{2} = \frac{d_{i,k} + d_{j,k} - d_{i,j}}{2}.$$

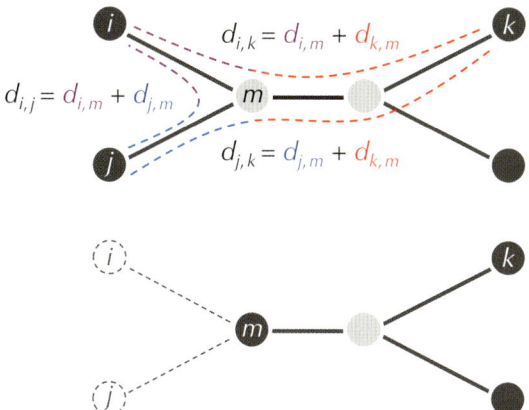

FIGURE 7.8 For neighboring leaves i and j and their parent node m, $d_{k,m} = (d_{i,k} + d_{j,k} - d_{i,j})/2$ for every other leaf k in the tree. (Bottom) Removing leaves i and j from the tree turns m into a leaf (we assume that m has degree 3). The distances from this new leaf to any other leaf k can be recomputed as $d_{k,m} = (D_{i,k} + D_{j,k} - D_{i,j})/2$.

Since i, j, and k are leaves, we can compute the distance $d_{k,m}$ between nodes k and m in terms of elements of the additive distance matrix D,

$$d_{k,m} = \frac{(D_{i,k} + D_{j,k} - D_{i,j})}{2}.$$

In the case when the parent m has degree 3 (as in Figure 7.8 (top)), removing leaves i and j from the tree turns m into a leaf and thus reduces the total number of leaves (Figure 7.8 (bottom)). This operation is equivalent to removing rows i and j as well as columns i and j from D, then adding a new row and column corresponding to their parent m, where the distances from m to other leaves are computed according to the above formula.

> **EXERCISE BREAK:** We have just described how to reduce the size of the tree as well as the dimension of the distance matrix D if the parent node (m) has degree 3. Design a similar approach in the case that the degree of m is larger than 3.

This discussion implies a recursive algorithm for the Distance-Based Phylogeny Problem:

- find a pair of neighboring leaves i and j by selecting the minimum element $D_{i,j}$ in the distance matrix;

CHAPTER 7

- replace *i* and *j* with their parent, and recompute the distances from this parent to all other leaves as described above;
- solve the Distance-Based Phylogeny problem for the smaller tree;
- add the previously removed leaves *i* and *j* back to the tree.

EXERCISE BREAK: Apply this recursive approach to the distance matrix shown in Figure 7.9 (left). (Solve this exercise by hand.)

	i	*j*	*k*	*l*			*i*	*j*	*k*	*l*
i	0	13	21	22		*i*	0	3	4	3
j	13	0	12	13		*j*	3	0	4	5
k	21	12	0	13		*k*	4	4	0	2
l	22	13	13	0		*l*	3	5	2	0

FIGURE 7.9 (Left) An additive 4 × 4 distance matrix. (Right) A non-additive 4 × 4 distance matrix.

Computing limb lengths

If you attempted the preceding exercise, then you were likely driven crazy. The reason why is that in the first step of our proposed algorithm, we assumed that a minimum element of an additive distance matrix corresponds to neighboring leaves. Yet as illustrated in Figure 7.10, this assumption is not necessarily true! Thus, we need a new approach to the Distance-Based Phylogeny Problem, as finding the animal coronavirus that is the smallest distance from SARS-CoV may not be the best way to identify the animal reservoir of SARS.

Our proposed recursive approach may have failed, but using recursion was a good idea, and so we will explore a different recursive algorithm. Rather than looking for a *pair* of neighbors in TREE(D), we will instead reduce the size of the tree by trimming its leaves *one at a time*. Of course, we don't know TREE(D), and so we must somehow trim leaves in TREE(D) by analyzing the distance matrix.

As a first step toward constructing TREE(D), we will address the more modest goal of computing the lengths of limbs in TREE(D). So, given a leaf *j* in a tree, we denote the length of the limb connecting *j* with its parent as LIMBLENGTH(j). Edges that are not limbs must connect two internal nodes and are therefore called **internal edges**.

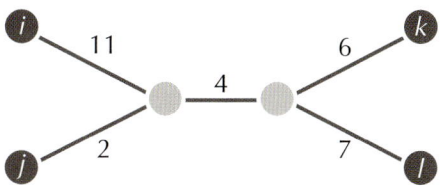

FIGURE 7.10 The simple tree fitting the distance matrix from Figure 7.9 (left). The two closest leaves in this tree (*j* and *k*) are not neighbors.

Limb Length Problem:

Compute the length of a limb in the simple tree fitting an additive distance matrix.

Input: An additive distance matrix D and an integer j.
Output: LIMBLENGTH(j), the length of the limb connecting leaf j to its parent in TREE(D).

To compute LIMBLENGTH(j) for a given leaf j, note that because TREE(D) is simple, we know that PARENT(j) has degree at least 3 (unless TREE(D) has only two nodes). We can therefore think of PARENT(j) as dividing the other nodes of TREE(D) into at least three **subtrees**, or smaller trees that would remain if we were to remove PARENT(j) along with any edges connecting it to other nodes (Figure 7.11). Because j is a leaf, it must belong to a subtree by itself; we call this subtree T_j. This brings us to the following result.

Limb Length Theorem: *Given an additive matrix D and a leaf j, LIMBLENGTH(j) is equal to the minimum value of $(D_{i,j} + D_{j,k} - D_{i,k})/2$ over all leaves i and k.*

Proof. A given pair of leaves can belong to the same subtree or to different subtrees. So first assume that leaves i and k belong to different subtrees T_i and T_k (Figure 7.11). Because PARENT(j) is on the path connecting i to k, it follows that

$$d_{i,j} = d_{i,\text{PARENT}(j)} + \text{LIMBLENGTH}(j)$$
$$d_{j,k} = d_{k,\text{PARENT}(j)} + \text{LIMBLENGTH}(j)$$

Adding these two equations yields

$$d_{i,j} + d_{j,k} = d_{i,\text{PARENT}(j)} + d_{k,\text{PARENT}(j)} + 2 \cdot \text{LIMBLENGTH}(j).$$

CHAPTER 7

Because $d_{i,\text{PARENT}(j)} + d_{k,\text{PARENT}(j)}$ is equal to $d_{i,k}$, it follows that

$$\text{LIMBLENGTH}(j) = \frac{d_{i,j} + d_{j,k} - d_{i,k}}{2} = \frac{D_{i,j} + D_{j,k} - D_{i,k}}{2}.$$

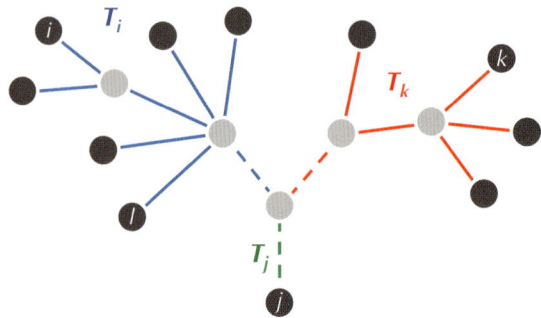

FIGURE 7.11 A simple tree with selected leaves i, j, k, and l. Removing the parent of j (along with the three dashed edges connecting it to other nodes) would separate this tree into three subtrees, whose edges are shown in different colors. Leaves i and l belong to T_i, whereas leaf k belongs to T_k. Leaf j belongs to T_j, which contains a single node.

On the other hand, assume that leaves i and l belong to the same subtree (Figure 7.11). Then the path from i to l does not pass through $\text{PARENT}(j)$, and so we have the *inequality*

$$d_{i,\text{PARENT}(j)} + d_{l,\text{PARENT}(j)} \geq d_{i,l}.$$

Combining this with the equation

$$d_{i,j} + d_{j,l} = d_{i,\text{PARENT}(j)} + d_{l,\text{PARENT}(j)} + 2 \cdot \text{LIMBLENGTH}(j)$$

yields that

$$\text{LIMBLENGTH}(j) = \frac{d_{i,j} + d_{j,l} - (d_{i,\text{PARENT}(j)} + d_{l,\text{PARENT}(j)})}{2}$$
$$\leq \frac{d_{i,j} + d_{j,l} - d_{i,l}}{2} = \frac{D_{i,j} + D_{j,l} - D_{i,l}}{2}.$$

As a result of this discussion, $\text{LIMBLENGTH}(j)$ must be less than or equal to $(D_{i,j} + D_{j,k} - D_{i,k})/2$ for any choice of leaves i and k. Because we can always find leaves i and k belonging to different subtrees (why?), it follows that $\text{LIMBLENGTH}(j)$ is equal to the minimum value of $(D_{i,j} + D_{j,k} - D_{i,k})/2$ over all choices of i and k. □

We now have an algorithm for solving the Limb Length Problem. For each j, we can compute LIMBLENGTH(j) by finding the minimum value of $(D_{i,j} + D_{j,k} - D_{i,k})/2$ over all pairs of leaves i and k.

EXERCISE BREAK: The proposed algorithm computes LIMBLENGTH(j) in $\mathcal{O}(n^2)$ time (for an $n \times n$ distance matrix). Design an algorithm that computes LIMBLENGTH(j) in $\mathcal{O}(n)$ time.

Additive Phylogeny

Trimming the tree

Since we now know how to find the length of any limb in TREE(D), we can construct TREE(D) recursively using the algorithm illustrated in Figure 7.12.

First, imagine that we already know TREE(D), and pick an arbitrary leaf j. We will trim the limb of j by reducing its length by LIMBLENGTH(j). Because we do not know TREE(D), we need to represent trimming the leaf j in terms of the distance matrix D. To do so, we first subtract LIMBLENGTH(j) from each off-diagonal element in row j and column j of D to obtain a matrix D^{bald} for which the limb of j has become a **bald limb**, or a limb of length 0 (Figure 7.12). We will further assume that a bald limb has disappeared from the tree entirely. In terms of the distance matrix, ignoring a bald limb means removing row j and column j from D to produce a smaller $(n-1) \times (n-1)$ distance matrix D^{trimmed}. We can now recursively find TREE(D) in four steps:

- pick an arbitrary leaf j, compute LIMBLENGTH(j), and construct the distance matrix D^{trimmed};
- solve the Distance-Based Phylogeny Problem for D^{trimmed};
- identify the point in TREE(D^{trimmed}) where leaf j should be attached in TREE(D);
- add a limb of length LIMBLENGTH(j) growing from this attachment point in TREE(D^{trimmed}) to form TREE(D).

STOP and Think: When adding leaf j back to TREE(D^{trimmed}), how would you find its attachment point?

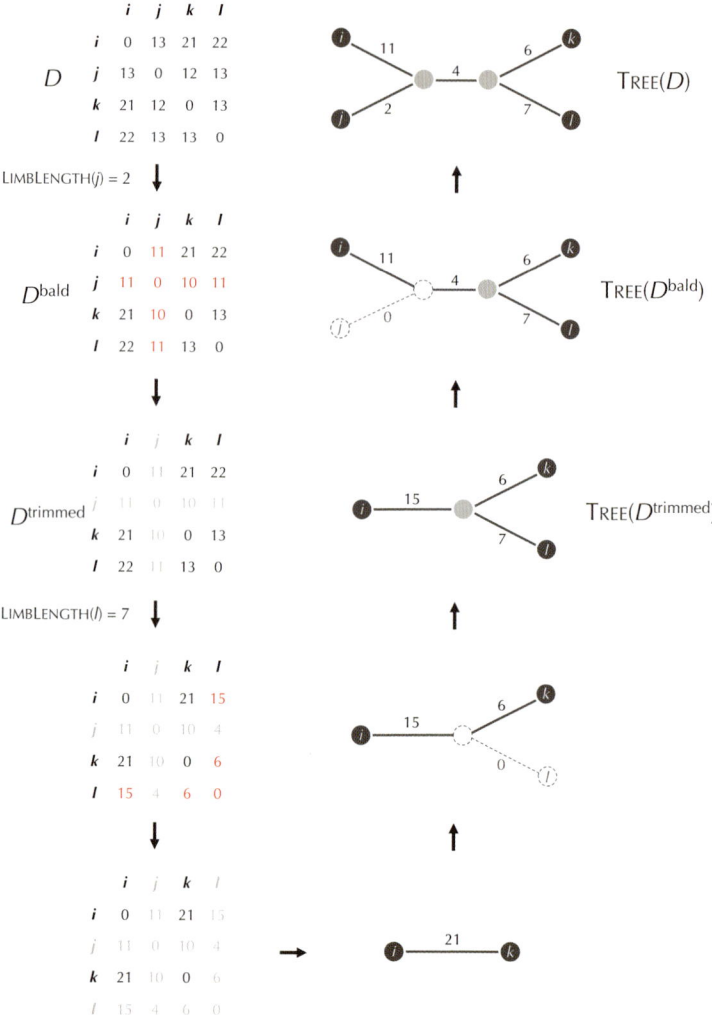

FIGURE 7.12 Converting the additive distance matrix from Figure 7.9 (left) into the simple tree fitting this matrix from Figure 7.10. On the left side, we first compute LIMBLENGTH(j) = 2, and then subtract 2 from the non-diagonal elements in row j and column j of D to obtain D^{bald} (updated values are shown in red). Removing this row and column yields a 3 × 3 distance matrix $D^{trimmed}$. We find that LIMBLENGTH(l) = 7 in $D^{trimmed}$ and subtract 7 from the non-diagonal elements in row l and column l. Graying out this row and column yields a 2 × 2 distance matrix. On the right side, we can fit this 2 × 2 distance matrix to a tree consisting of a single edge. By finding the attachment points of removed limbs (shown on the left), we reconstruct TREE($D^{trimmed}$), TREE(D^{bald}), and then TREE(D).

Attaching a limb

To find the attachment point of a leaf j in TREE(D^{trimmed}), consider TREE(D^{bald}), which is the same as TREE(D) except that LIMBLENGTH(j) = 0. From the Limb Length Theorem, we know that there must be leaves i and k in TREE(D^{bald}) such that

$$\frac{D^{\text{bald}}_{i,j} + D^{\text{bald}}_{j,k} - D^{\text{bald}}_{i,k}}{2} = 0,$$

which implies that

$$D^{\text{bald}}_{i,k} = D^{\text{bald}}_{i,j} + D^{\text{bald}}_{j,k}.$$

Thus, the attachment point for leaf j must be located at distance $D^{\text{bald}}_{i,j}$ from leaf i on the path connecting i and k in the trimmed tree. This attachment point may occur at an existing node, in which case we connect j to this node. On the other hand, the attachment point for j may occur along an edge, in which case we place a new node at the attachment point and connect j to it.

An algorithm for distance-based phylogeny construction

The preceding discussion results in the recursive algorithm below, which we call **ADDITIVEPHYLOGENY**, for finding the simple tree fitting an $n \times n$ additive distance matrix D. We assume that you have already implemented a program **LIMB**(D, j) that computes LIMBLENGTH(j) for a leaf j based on the distance matrix D. Rather than selecting an arbitrary leaf j from TREE(D) for trimming, **ADDITIVEPHYLOGENY** selects leaf n (corresponding to the last row and column of D).

> **STOP and Think:** Consider these questions about **ADDITIVEPHYLOGENY**.
>
> - What is its running time?
>
> - Although it may seem that **ADDITIVEPHYLOGENY** would construct a tree for any matrix, this is not the case. What goes wrong if you apply **ADDITIVEPHYLOGENY** to the non-additive distance matrix in Figure 7.9 (right)?
>
> - Modify **ADDITIVEPHYLOGENY** to develop an algorithm that checks whether a given distance matrix is additive. Then, apply this test to the distance matrix for coronavirus Spike proteins shown in Figure 7.13 (top). Is this matrix additive?

CHAPTER 7

> **ADDITIVEPHYLOGENY**(D, n)
>   **if** n = 2
>   **return** the tree consisting of a single edge of length $D_{1,2}$
>   limbLength ← **LIMB**(D, n)
>   **for** j ← 1 to n − 1
> $D_{j,n}$ ← $D_{j,n}$ − limbLength
> $D_{n,j}$ ← $D_{j,n}$
>   (i, n, k) ← three leaves such that $D_{i,k} = D_{i,n} + D_{n,k}$
>   x ← $D_{i,n}$
>   remove row n and column n from D
>   T ← **ADDITIVEPHYLOGENY**(D, n − 1)
>   v ← the (potentially new) node in T at distance x from i on the path between i and k
>   add leaf n back to T by creating a limb (v, n) of length limbLength
>   **return** T

Although the previous question suggests that **ADDITIVEPHYLOGENY** can be modified to determine whether a given distance matrix is additive, there exists an even simpler way to check for additivity (see **DETOUR: The Four Point Condition**).

Constructing an evolutionary tree of coronaviruses

By the end of 2003, bioinformaticians had sequenced many coronaviruses from a variety of birds and mammals, from which we obtain the distance matrix in Figure 7.13 (top) based on a multiple alignment of Spike proteins.

Although you now understand the perils of concluding that the minimum element of the distance matrix corresponds to a pair of neighbors, common sense tells us with a glance at Figure 7.13 (top) that the civet must be the animal reservoir of SARS. This information led researchers to hypothesize that inadequate preparation of meat from palm civets (Figure 7.14) in the Guangdong region of China may have caused the SARS outbreak.

Yet before rushing to this conclusion, you may like to read **DETOUR: Did Bats Give Us SARS?** to see why the history of interspecies viral transfer is often difficult to trace. In fact, some studies have suggested that humans first received SARS from bats, which later gave the virus to palm civets, which then transmitted the disease back to humans. The civet was identified as the animal reservoir of SARS in 2003 in part because SARS viruses from other potential suspects, including bats, had not yet been sequenced.

WHICH ANIMAL GAVE US SARS?

STOP and Think: It turns out that most distance matrices constructed from real data (including the distance matrix in Figure 7.13 (top)) are non-additive. Why do you think that this is the case?

Since the distance matrix for SARS-like coronaviruses is non-additive, we will cheat a bit and slightly modify it to make it additive so that you can apply **ADDITIVEPHYLOGENY** to it (Figure 7.13 (bottom)).

	Cow	Pig	Horse	Mouse	Dog	Cat	Turkey	Civet	Human
Cow	0	295	300	524	1077	1080	978	941	940
Pig	295	0	314	487	1071	1088	1010	963	966
Horse	300	314	0	472	1085	1088	1025	965	956
Mouse	524	487	472	0	1101	1099	1021	962	965
Dog	1076	1070	1085	1101	0	818	1053	1057	1054
Cat	1082	1088	1088	1098	818	0	1070	1085	1080
Turkey	976	1011	1025	1021	1053	1070	0	963	961
Civet	941	963	965	962	1057	1085	963	0	16
Human	940	966	956	965	1054	1080	961	16	0

	Cow	Pig	Horse	Mouse	Dog	Cat	Turkey	Civet	Human
Cow	0	295	306	497	1081	1091	1003	956	954
Pig	295	0	309	500	1084	1094	1006	959	957
Horse	306	309	0	489	1073	1083	995	948	946
Mouse	497	500	489	0	1092	1102	1014	967	965
Dog	1081	1084	1073	1092	0	818	1056	1053	1051
Cat	1091	1094	1083	1102	818	0	1066	1063	1061
Turkey	1003	1006	995	1014	1056	1066	0	975	973
Civet	956	959	948	967	1053	1063	975	0	16
Human	954	957	946	965	1051	1061	973	16	0

FIGURE 7.13 (Top) The distance matrix based on pairwise alignment of Spike proteins from coronaviruses extracted from various animals. The distance between each pair of sequences was computed as the total number of mismatches and indels in their optimal alignment. (Bottom) A modification of the distance matrix to make it additive.

EXERCISE BREAK: Construct the simple tree fitting the distance matrix in Figure 7.13 (bottom).

CHAPTER 7

FIGURE 7.14 The palm civet.

Using Least Squares to Construct Approximate Distance-Based Phylogenies

If an $n \times n$ distance matrix D is non-additive, then we will instead look for a weighted tree T whose distances between leaves approximate the entries in D. To this end, we would like for T to minimize the **sum of squared errors** DISCREPANCY(T, D), which is given by the formula

$$\text{DISCREPANCY}(T, D) = \sum_{1 \leq i < j \leq n} (d_{i,j}(T) - D_{i,j})^2 \,.$$

Least Squares Distance-Based Phylogeny Problem:

Given a distance matrix, find the tree that minimizes the sum of squared errors.

Input: An $n \times n$ distance matrix D.
Output: A weighted tree T minimizing DISCREPANCY(T, D) over all weighted trees with n leaves.

EXERCISE BREAK: Let T be the tree in Figure 7.10 with all edge lengths removed. Given the non-additive 4×4 distance matrix D in Figure 7.9 (right), find the lengths of edges in this tree that minimize DISCREPANCY(T, D).

It turns out that for a specific tree T, it is easy to find edge weights in T minimizing DISCREPANCY(T, D). Yet our ability to minimize the sum of squared errors for a *specific* tree does not imply that we can efficiently solve the Least Squares Distance-Based Phylogeny Problem, since the number of different trees grows very quickly as the number of leaves in the tree increases. In fact, the Least Squares Distance-Based Phylogeny winds up being *NP*-Complete, and so we must abandon the hope of designing a fast algorithm to find a tree that best fits a non-additive matrix. In the next two sections, we will explore heuristics for constructing trees from non-additive matrices that solve this problem approximately.

Ultrametric Evolutionary Trees

Biologists often assume that every internal node in an evolutionary tree corresponds to a species that underwent a **speciation event**, splitting one ancestral species into two descendants. Note that every internal node in the tree in Figure 7.15 (top) (corresponding to a speciation event) has degree 3. We therefore define an **unrooted binary tree** as a tree where every node has degree equal to either 1 or 3.

> **EXERCISE BREAK:** Prove that every unrooted binary tree with n leaves has $n-2$ internal nodes (and thus $2n-3$ edges).

A **rooted binary tree** is an unrooted binary tree that has a root (of degree 2) placed on one of its edges. In other words, we replace an edge (v, w) with a root and draw edges connecting the root to each of v and w (Figure 7.15 (bottom)).

If we had a **molecular clock** measuring evolutionary time, then we could assign an **age** to every node v in a rooted binary tree (denoted $\text{AGE}(v)$), where all of the leaves of the tree have age 0 because they correspond to present-day species. We could then define the weight of an edge (v, w) in the tree as the difference $\text{AGE}(v) - \text{AGE}(w)$. Consequently, the length of a path between the root and any node would be equal to the difference between their ages. Such a tree, in which the distance from the root to any leaf is the same, is called **ultrametric** (Figure 7.16 (bottom right)).

Our aim is to derive an ultrametric tree that explains a given distance matrix (even if it does so only approximately). **UPGMA** (which stands for **U**nweighted **P**air **G**roup **M**ethod with **A**rithmetic Mean) is a simple clustering heuristic that introduces a hypothetical molecular clock for constructing an ultrametric evolutionary tree. You can learn more about clustering in Chapter 8.

Given an $n \times n$ matrix D, **UPGMA** (which is illustrated in Figure 7.16) first forms n trivial clusters, each containing a single leaf. The algorithm then finds a pair of "closest" clusters. To clarify the notion of closest clusters, **UPGMA** defines the distance between clusters C_1 and C_2 as the average pairwise distance between elements of C_1 and C_2,

$$D_{C_1, C_2} = \frac{\sum_{i \in C_1} \sum_{j \in C_2} D_{i,j}}{|C_1| \cdot |C_2|}.$$

CHAPTER 7

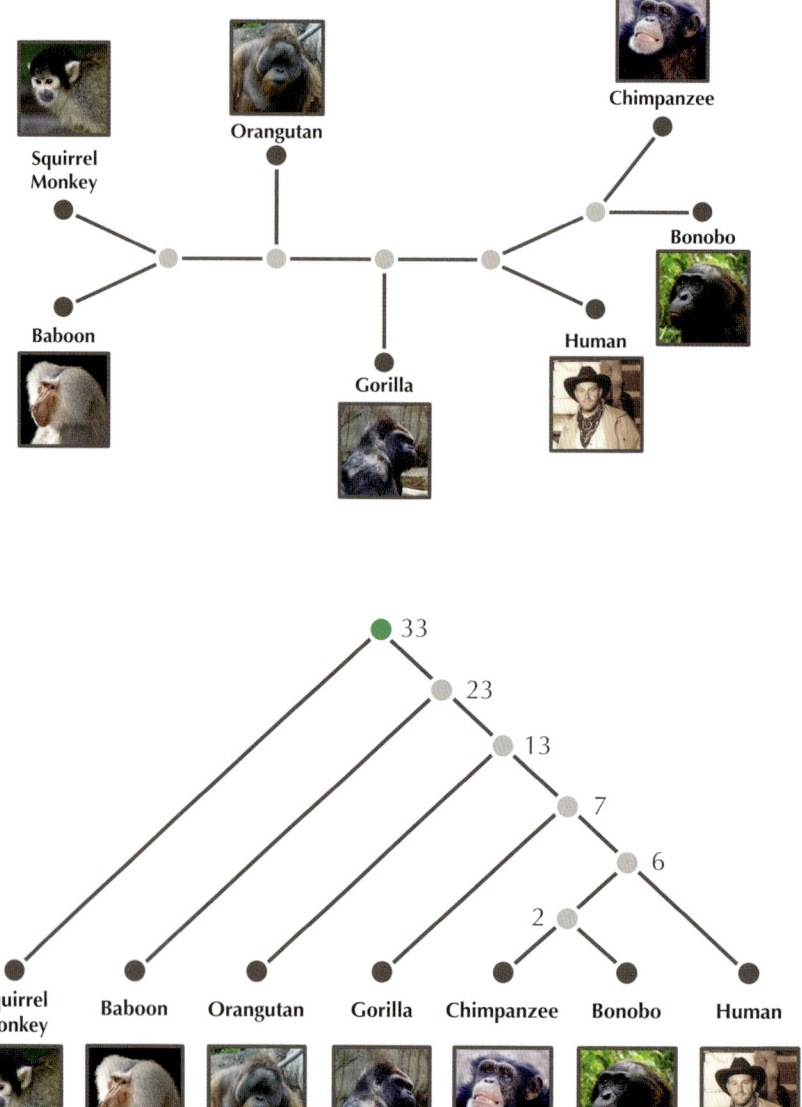

FIGURE 7.15 (Top) An unrooted binary tree representing a phylogeny of primates. (Bottom) Placing a root on the squirrel monkey's limb results in a rooted binary tree. The number at each node corresponds to the number of million years ago that the divergence at this node occurred.

WHICH ANIMAL GAVE US SARS?

In this equation, the notation $|C|$ denotes the number of leaves in cluster C.

Once **UPGMA** has identified a pair of closest clusters C_1 and C_2, it **merges** them into a cluster C with $|C_1| + |C_2|$ elements and then creates a node for C, which it connects to each of C_1 and C_2 by a directed edge. The age of C is set to be $D_{C_1,C_2}/2$. **UPGMA** then iterates this process of merging the two closest clusters until only a single cluster remains, which corresponds to the root.

UPGMA(*D*, *n*)
   Clusters ← *n* single-element clusters labeled 1, ..., *n*
   construct a graph *T* with *n* isolated nodes labeled by single elements 1, ..., *n*
   for every node *v* in *T*
      AGE(*v*) ← 0
   while there is more than one cluster
      find the two closest clusters C_i and C_j (break ties arbitrarily)
      merge C_i and C_j into a new cluster C_{new} with $|C_i| + |C_j|$ elements
      add a new node labeled by cluster C_{new} to *T*
      connect node C_{new} to C_i and C_j by directed edges
      AGE(*C*) ← $D_{C_i,C_j}/2$
      remove the rows and columns of *D* corresponding to C_i and C_j
      remove C_i and C_j from *Clusters*
      add a row/column to *D* for C_{new} by computing $D(C_{new}, C)$ for each *C* in *Clusters*
      add C_{new} to *Clusters*
   root ← the node in *T* corresponding to the remaining cluster
   for each edge (*v*, *w*) in *T*
      length of (*v*, *w*) ← AGE(*v*) − AGE(*w*)
   return *T*

EXERCISE BREAK: Prove that after merging clusters C_i and C_j into a cluster C_{new}, the distance between C_{new} and another cluster C_m is equal to $(D_{C_i,C_m} \cdot |C_i| + D_{C_j,C_m} \cdot |C_j|) / (|C_i| + |C_j|)$.

UPGMA offers a step forward from **ADDITIVEPHYLOGENY**, since it can analyze non-additive distance matrices. Figure 7.17 shows the result of applying UPGMA to the coronavirus distance matrix from Figure 7.13 (top). However, the first step that **UPGMA** takes is to merge the two leaves *i* and *j* with minimum distance $D_{i,j}$ into a single cluster. And we have already seen that the smallest element in the distance matrix does

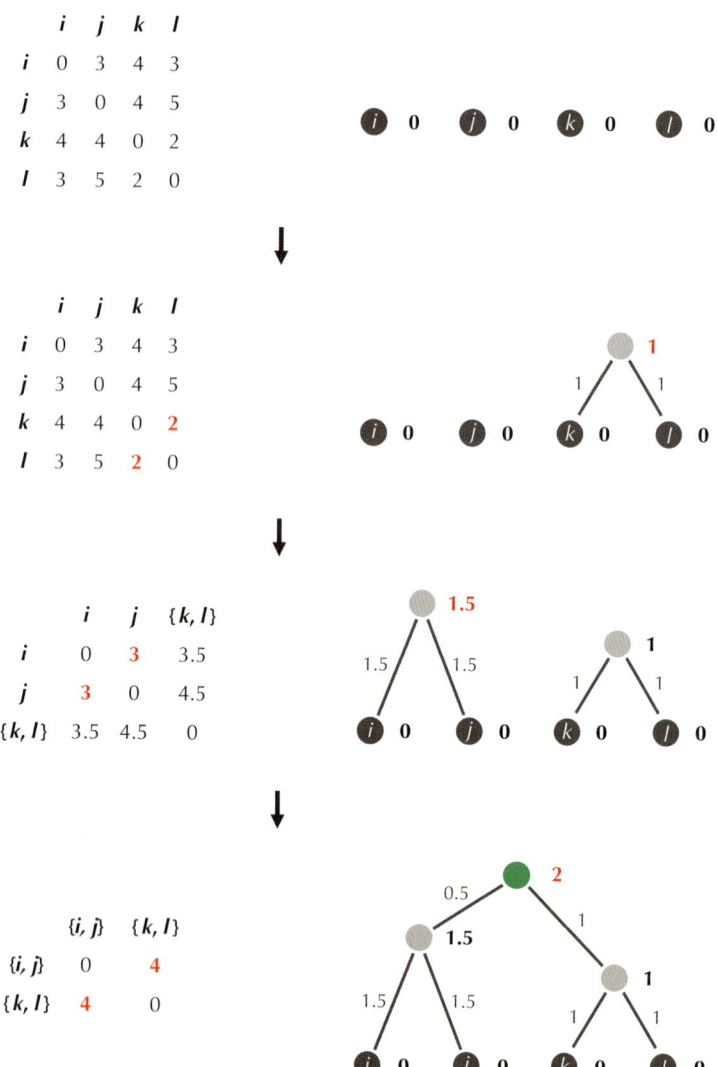

FIGURE 7.16 Tree reconstruction with **UPGMA** for the non-additive distance matrix from Figure 7.9 (right). **UPGMA** begins with forming one cluster for each leaf. In each step, it identifies the two closest clusters C_1 and C_2, merge them into a new node C, and connect C to C_1 and C_2 by directed edges. The age of C is set equal to $D_{C_1,C_2}/2$. We then iterate this process until only a single cluster remains, which must be the root. The resulting tree is ultrametric (i.e., the distance from the root to any leaf is the same).

not necessarily correspond to a pair of neighboring leaves! This is a concern, since if **UPGMA** generates incorrect trees from additive matrices, then it is not an ideal heuristic for evolutionary tree construction from non-additive matrices. Can we find an algorithm that always identifies neighboring leaves in an additive distance matrix but also performs well on a non-additive distance matrix?

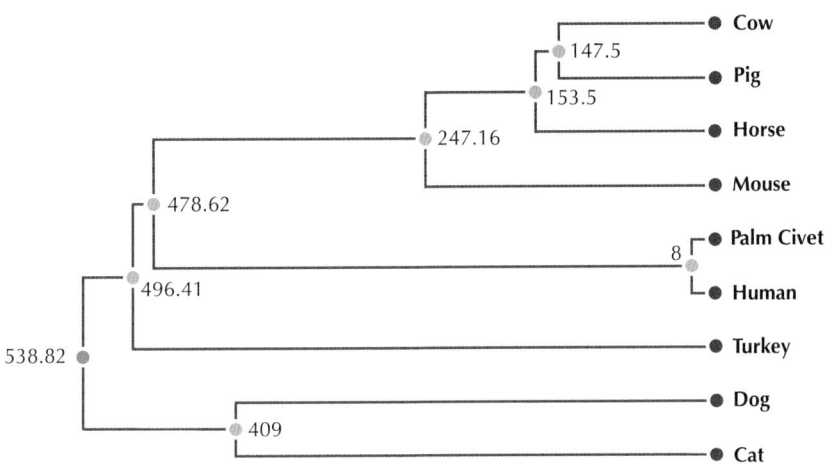

FIGURE 7.17 The ultrametric tree of coronaviruses created by **UPGMA** using the distance matrix in Figure 7.13 (top). The root is shown in green.

The Neighbor-Joining Algorithm

Transforming a distance matrix into a neighbor-joining matrix

In 1987, Naruya Saitou and Masatoshi Nei developed the **neighbor-joining algorithm** for evolutionary tree reconstruction. Given an additive distance matrix, this algorithm, which we call **NEIGHBORJOINING**, finds a pair of neighboring leaves and substitutes them by a single leaf, thus reducing the size of the tree. **NEIGHBORJOINING** can thus recursively construct a tree fitting the additive matrix. This algorithm also provides a heuristic for non-additive distance matrices that performs well in practice.

The central idea of **NEIGHBORJOINING** is that although finding a minimum element in a distance matrix D is not guaranteed to yield a pair of neighbors in TREE(D), we can transform D into a different matrix whose minimum element does yield a pair of neighbors. First, given an $n \times n$ distance matrix D, we define $\text{TOTALDISTANCE}_D(i)$ as the sum $\sum_{1 \leq k \leq n} D_{i,k}$ of distances from leaf i to all other leaves. The **neighbor-joining matrix** D^* is defined such that for all i and j, $D^*_{i,i} = 0$ and

$$D^*_{i,j} = (n-2) \cdot D_{i,j} - \text{TOTALDISTANCE}_D(i) - \text{TOTALDISTANCE}_D(j).$$

NEIGHBORJOINING, which is illustrated in Figure 7.18, is a widely used method for evolutionary tree reconstruction; the paper that introduced it is one of the most cited in all of science, with over 30,000 citations. Yet this algorithm is non-intuitive: the above formula for computing the matrix D^* probably looks like witchcraft to you. In fact, despite having flawless intuition, Saitou and Nei *never proved* that their algorithm correctly solves the Distance-Based Phylogeny Problem for additive matrices! However, it took researchers another year to prove the following theorem, whose proof we have passed to DETOUR: Why Does the Neighbor-Joining Algorithm Find Neighboring Leaves?

Neighbor-Joining Theorem: *Given an additive matrix D, the smallest element $D^*_{i,j}$ of its neighbor-joining matrix D^* corresponds to a pair of neighboring leaves i and j in $\text{TREE}(D)$.*

If $n = 2$, then **NEIGHBORJOINING**(D, n) returns the tree consisting of a single edge of length $D_{1,2}$. If $n > 2$, then it selects the minimum element in the neighbor-joining matrix, replaces the neighboring leaves i and j with a new leaf m, and then computes the distance from m to any other leaf k according to the formula

$$D_{k,m} = \tfrac{1}{2}(D_{k,i} + D_{k,j} - D_{i,j}),$$

which is motivated by Figure 7.8. This equation allows us to replace an $n \times n$ matrix D with an $(n-1) \times (n-1)$ matrix D' in which i and j have been replaced by m. By recursively applying **NEIGHBORJOINING** to D', we obtain an evolutionary tree with $n-1$ leaves. We then add two limbs starting at node m, one ending in leaf i and the other ending in leaf j. We set

$$\Delta_{i,j} = \frac{\text{TOTALDISTANCE}_D(i) - \text{TOTALDISTANCE}_D(j)}{n-2}$$

and assign

$$\text{LIMBLENGTH}(i) = \frac{1}{2}\left(D_{i,j} + \Delta_{i,j}\right)$$

$$\text{LIMBLENGTH}(j) = \frac{1}{2}\left(D_{i,j} - \Delta_{i,j}\right)$$

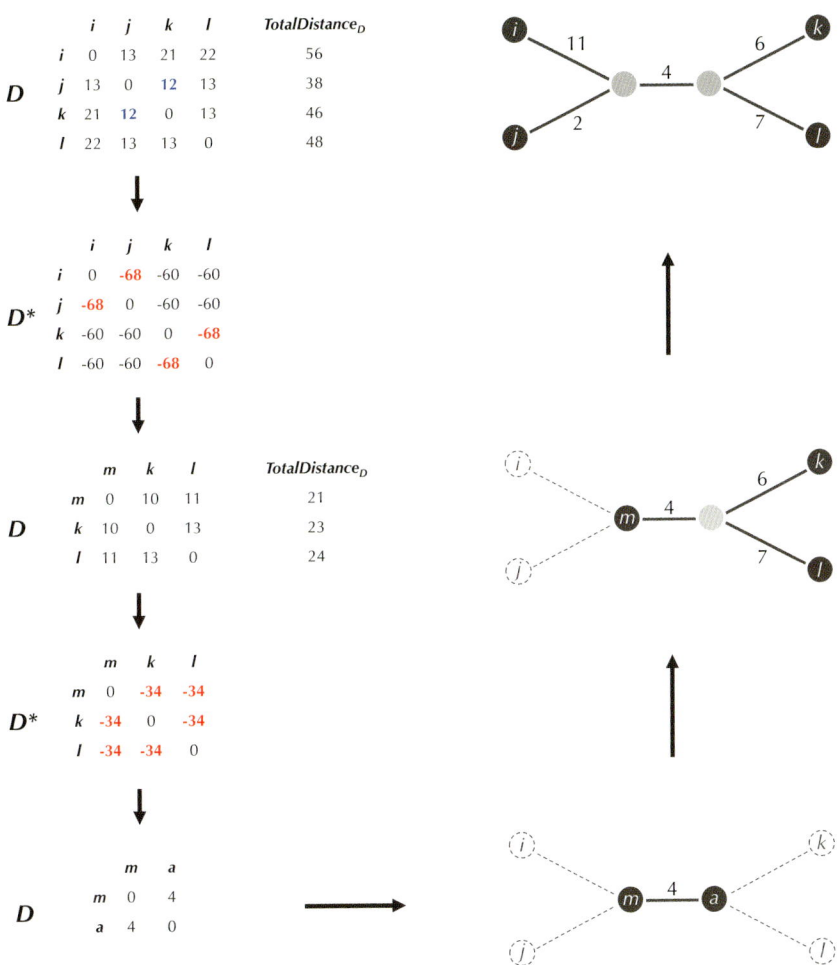

FIGURE 7.18 (Top left) The additive distance matrix D from Figure 7.9 (left) along with the array $TotalDistance_D$. $D_{j,k}$ (shown in blue) is a minimum element of D, but as it turns out, leaves j and k are not neighbors in $\text{TREE}(D)$. Moving down the left side, we construct the neighbor-joining matrix D^* and find that $D^*_{i,j}$ (red) is a minimum element of D^*. We transform the initial 4×4 distance matrix into a 3×3 distance matrix by replacing i and j with a single leaf m and updating distances from m to other leaves as $D_{k,m} = \frac{1}{2}(D_{k,i} + D_{k,j} - D_{i,j}) = \frac{1}{2}(21 + 12 - 13) = 10$ and $D_{l,m} = \frac{1}{2}(D_{l,i} + D_{l,j} - D_{i,j}) = \frac{1}{2}(22 + 13 - 13) = 11$. We select $D^*_{k,l}$ as a minimum element of D^* and replace leaves k and l with a single leaf a. The resulting 2×2 matrix corresponds to a tree with a single edge connecting m and a. We then work our way up the right side, adding pairs of neighbors back into the tree at each step using formulas for limb lengths. (Top right) The tree $\text{TREE}(D)$ fitting the original matrix D.

CHAPTER 7

To see where these formulas come from, see **DETOUR: Computing Limb Lengths in the Neighbor-Joining Algorithm**.

EXERCISE BREAK: Prove that if D is additive, then for any i and j between 1 and n, both $\frac{1}{2}(D_{i,j} + \Delta_{i,j})$ and $\frac{1}{2}(D_{i,j} - \Delta_{i,j})$ are non-negative.

The following pseudocode summarizes the neighbor-joining algorithm.

NEIGHBORJOINING(D, n)
    if $n = 2$
        $T \leftarrow$ the tree consisting of a single edge of length $D_{1,2}$
        return T
    $D^* \leftarrow$ the neighbor-joining matrix constructed from the distance matrix D
    find elements i and j such that $D^*_{i,j}$ is a minimum non-diagonal element of D^*
    $\Delta \leftarrow (\text{TOTALDISTANCE}_D(i) - \text{TOTALDISTANCE}_D(j))/(n-2)$
    $limbLength_i \leftarrow \frac{1}{2}(D_{i,j} + \Delta)$
    $limbLength_j \leftarrow \frac{1}{2}(D_{i,j} - \Delta)$
    add a new row/column m to D so that $D_{k,m} = D_{m,k} = \frac{1}{2}(D_{k,i} + D_{k,j} - D_{i,j})$
        for any k
    remove rows i and j from D
    remove columns i and j from D
    $T \leftarrow$ **NEIGHBORJOINING**(D, $n-1$)
    add two new limbs (connecting node m with leaves i and j) to the tree T
    assign length $limbLength_i$ to $\text{LIMB}(i)$
    assign length $limbLength_j$ to $\text{LIMB}(j)$
    return T

EXERCISE BREAK: Before implementing **NEIGHBORJOINING**, apply it to the additive and non-additive distance matrices from Figure 7.9.

EXERCISE BREAK: Apply **NEIGHBORJOINING** to the coronavirus distance matrix from Figure 7.13 (top).

Analyzing coronaviruses with the neighbor-joining algorithm

Figure 7.19 shows the neighbor-joining tree of coronaviruses isolated from different animals based on the distance matrix in Figure 7.13 (top). We can also apply **NeighborJoining** to the distance matrix of SARS-CoV variants isolated from various human carriers, in addition to a coronavirus from palm civet (Figure 7.20).

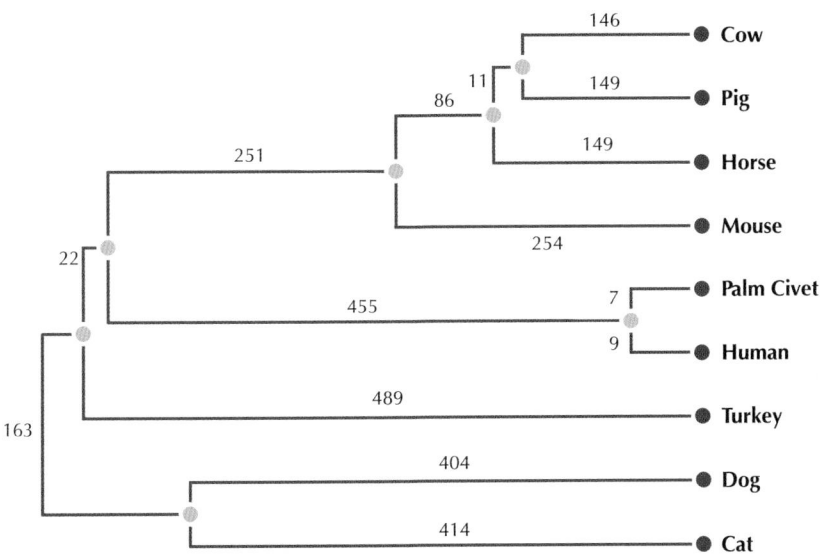

FIGURE 7.19 The neighbor-joining tree of coronaviruses isolated from different animals, based on the non-additive distance matrix in Figure 7.13 (top).

The SARS-CoV strain labeled "Hanoi" in Figure 7.20 was taken from Carlo Urbani, an Italian physician who worked for the World Health Organization. In February 2003, Urbani was called into a Hanoi hospital to examine a patient who had fallen ill with what local doctors believed to be a bad case of influenza; these doctors were afraid that it might be bird flu. In fact, the patient was the American man who had stayed across the hall from Liu Jianlun at the Metropole Hotel just one week earlier.

Fortunately, Urbani was quick to realize that the disease was not influenza, and he became the first physician to raise the alarm to public health officials. Yet rather than leaving Hanoi, he demanded that he remain there in order to oversee quarantine procedures. In an argument with his wife, who scolded him for risking his life to treat sick patients. Urbani replied, "What am I here for? Answering e-mails, going to cocktail parties and pushing paper?" Urbani would ultimately lose his life to SARS a month

CHAPTER 7

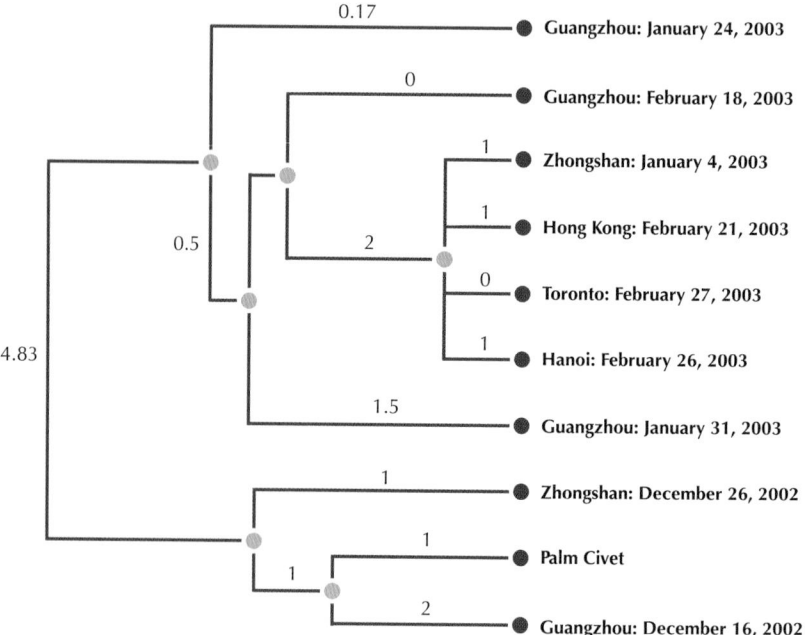

	Guangzhou Dec. 16, 2002	Zhongshan Dec. 16, 2002	Guangzhou Jan. 24, 2003	Guangzhou Jan. 31, 2003	Guangzhou Feb. 18, 2003	Hong Kong Feb. 18, 2003	Hanoi Feb. 26, 2003	Toronto Feb. 17, 2003	Hong Kong Mar. 15, 2003	Palm Civet
Guangzhou	0	4	12	8	9	9	12	12	11	3
Zhongshan	4	0	10	6	7	7	10	10	9	3
Guangzhou	12	10	0	4	5	3	2	2	1	11
Guangzhou	8	6	4	0	3	1	4	4	3	7
Guangzhou	9	7	5	3	0	2	5	5	4	8
Hong Kong	9	7	3	1	2	0	3	3	2	8
Hanoi	12	10	2	4	5	3	0	2	1	11
Toronto	12	10	2	4	5	3	2	0	1	11
Hong Kong	11	9	1	3	4	2	1	1	0	10
Palm Civet	3	3	11	7	8	8	11	11	10	0

FIGURE 7.20 (Top) The distance matrix based on pairwise alignment of Spike proteins from SARS-CoV strains extracted from various patients as well as a coronavirus taken from a palm civet. The distance between each pair of sequences was computed as the total number of mismatches and indels in their optimal pairwise alignment. (Bottom) The evolutionary tree of these viruses constructed by the neighbor-joining algorithm.

later. But his sacrifice also helped begin the massive worldwide action against this disease that may well have saved millions of lives.

Limitations of distance-based approaches to evolutionary tree construction

Although distance-based tree reconstruction has successfully resolved questions about the origin and spread of SARS, many evolutionary controversies cannot be resolved by using a distance matrix. For example, when we convert each pair of rows of a multiple alignment into a distance value, we lose information contained in the alignment. As a result, distance-based methods do not allow us to reconstruct ancestral sequences of Spike proteins (corresponding to internal nodes in Figure 7.20), which could lead us to believe that such molecular paleontology is impossible. Therefore, a superior approach to evolutionary tree reconstruction would be to somehow use the alignment directly, without first converting it to a distance matrix.

Character-Based Tree Reconstruction

Character tables

Fifty years ago, biologists constructed phylogenies not from DNA and protein sequences but from anatomical and physiological features called **characters**. For example, when analyzing invertebrate evolution, one commonly used character is the presence or absence of wings, while another is the number of legs (varying from 0 to over 300 in some centipedes). These two characters result in the 3×2 **character table** shown in Figure 7.21 for three species.

In general, every row in an $n \times m$ character table represents a **character vector**, holding the values of m characters corresponding to one of n existing species. Our goal, roughly stated, is to construct an evolutionary tree in which leaves corresponding to present-day species with similar character vectors occur near each other in the tree. We would also like to assign m character values to each internal node in the tree in order to best explain the characters of ancestral species.

STOP and Think: Can you transform the preceding vague description into a well-formulated computational problem that models tree reconstruction based on character tables?

CHAPTER 7

	wings	legs
winged stick insect	Yes	6
wingless stick insect	No	6
giant centipede	No	42

FIGURE 7.21 (Top panel) Winged (left) and wingless (middle) stick insects, each having six legs, and the giant centipede (right), which has 42 legs. (Bottom panel) A 3 × 2 character table describes two characters (wings and legs) in these three invertebrates.

From anatomical to genetic characters

In 1965, Emile Zuckerkandl and Linus Pauling published "Molecules as documents of evolutionary history", arguing that DNA sequences offer a much more informative source of data than anatomical and physiological characters. The idea may seem obvious today, especially after we have spent half of a chapter constructing evolutionary trees from distance matrices generated from DNA sequences. However, Zuckerkandl and Pauling's proposal was initially met with skepticism by many biologists, who felt that DNA analysis could not offer the same power as anatomical comparison. A now-famous argument was initiated when Zuckerkandl and Pauling found that the amino acid sequence of human beta-hemoglobin is very similar to that of gorillas, prompting Zuckerkandl to write in 1963:

From the point of view of hemoglobin structure, it appears that gorilla is just an abnormal human.

Yet the surprising similarity between hemoglobin proteins in various primates flew in the face of clear anatomical differences between primates. As a result, the leading evolutionary biologist Gaylord Simpson immediately responded to Zuckerkandl:

...that is of course nonsense. What the comparison really indicates is that hemoglobin is a bad choice and has nothing to tell us about attributes, or indeed tells us a lie.

Despite such vehement initial criticisms, genetic analysis had become the dominant technique in evolutionary studies by the 1970s. Indeed, the analysis of DNA sequences answered evolutionary questions that previous analysis of anatomical characters had failed to resolve. Early examples include the classification of the giant panda (see **DETOUR: Giant Panda: Bear or Raccoon?**) and the identification of human origins (see **DETOUR: Where Did Humans Come From?**). As a result, most evolutionary experts had no choice but to adapt their views and become experts on molecular evolution.

Ironically, modern evolutionary studies often view an $n \times m$ multiple alignment as an $n \times m$ character table, with each column representing a character of its own. Our goal is to construct a tree whose leaves correspond to the rows of this alignment and whose internal nodes correspond to ancestral sequences in accordance with the most parsimonious evolutionary scenario. Before rigorously defining a most parsimonious scenario, we will describe one example of how this algorithmic framework solved a longstanding puzzle in insect evolution.

How many times has evolution invented insect wings?

Wings provided a revolutionary adaptation for insects, allowing them to escape predators and disperse into new territories, thus leading to many new insect species. Yet despite the evolutionary advantages provided by wings, some insects are apparently better equipped for survival without them. In fact, nearly all winged species are related to many wingless counterparts belonging to the same genus, and some entire orders of insects are wingless, including fleas and lice.

The acquisition of wings would seem to pose an evolutionary challenge because complex physiological interactions are required to accommodate flight. As a result, we would be led to believe that wings evolved only once in insects. This argument parallels **Dollo's principle of irreversibility**, a hypothesis proposed by 19th Century paleontologist Louis Dollo. According to this principle, when a species loses a complex organ, such as wings, the organ will not reappear in exactly the same form in the species's descendants.

STOP and Think: What do you think about this argument as it pertains to insect wings?

Until recently, biologists followed Dollo's principle with respect to insect wings, believing that re-evolution of wings was essentially impossible because unused flight genes in wingless insects would be free to accumulate mutations, eventually eroding into non-functional pseudogenes. However, in 2003, Michael Whiting studied various winged and wingless stick insects from around the world and refuted this argument. He sequenced an approximately 2,000 nucleotide-long segment (the **18S ribosomal RNA** gene) from these stick insects and constructed an evolutionary tree based on these sequences. From this phylogeny, he inferred that wings were re-invented at least three times and lost at least four times during stick insect evolution (Figure 7.22).

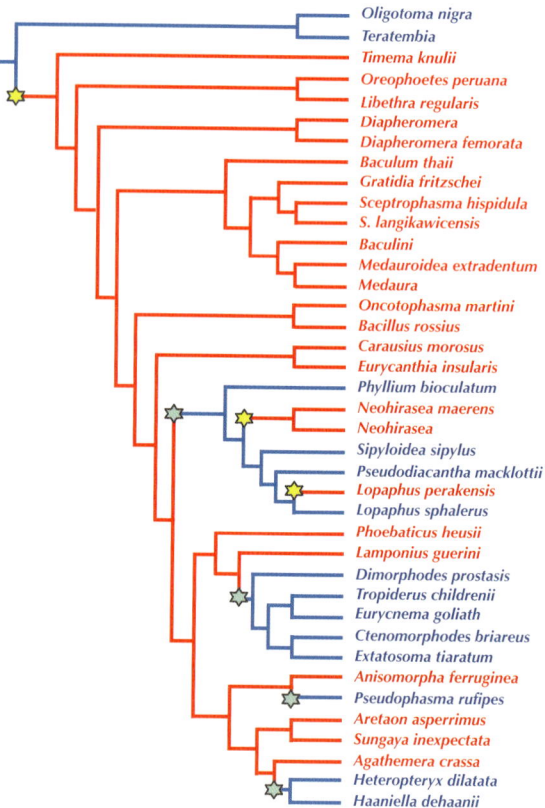

FIGURE 7.22 Evolutionary tree of winged (blue) and wingless (red) stick insects constructed from 18S ribosomal RNA genes. Transitions from winged to wingless species are shown by yellow stars; transitions from wingless to winged species are shown by green stars. 18S ribosomal RNAs are slow-evolving, making them ideal for reconstructing ancient divergences.

STOP and Think: Is it possible to infer an evolutionary scenario from the tree in Figure 7.22 that re-invents wings fewer than four times?

Whiting's work indicates the inherent complications of trying to infer evolutionary trees from anatomical characters. If you were asked to design a character-based phylogeny algorithm for the collection of all stick insects, then one of the first steps you would probably take would be to cluster all the winged insects on the opposite side of the tree from all of the wingless insects. However, the re-evolution of insect wings means that such an approach is flawed. Anatomical characters have even less power when we move into the microscopic world: just imagine trying to construct a phylogeny from direct observation of coronaviruses!

The Small Parsimony Problem

We will label each leaf of a tree by a row of a multiple alignment, and we will attempt to infer strings labeling internal nodes and corresponding to candidate ancestral sequences. However, we will need to develop a scoring function in order to quantify how well such a labeled tree fits the given multiple alignment. In what follows, we assume for simplicity that the multiple alignment contains only substitutions and no indels. In practice, researchers may start from a multiple alignment containing indels and then remove all columns containing indels.

An intuitive score of an evolutionary tree is the total number of mutations required to explain the strings at all nodes of the tree. Given a tree T with every node labeled by a string of length m, we will therefore set the length of edge (v, w) equal to the number of substitutions (Hamming distance) between the strings labeling v and w. The **parsimony score** of T is the sum of the lengths of its edges (Figure 7.23).

We will first assume that we are given the structure of a rooted binary tree in advance, in which case we only need to assign strings to the internal nodes in order to minimize the parsimony score.

Small Parsimony Problem:
Find the most parsimonious labeling of the internal nodes of a rooted tree.

> **Input**: A rooted binary tree with each leaf labeled by a string of length m.
> **Output**: A labeling of all other nodes of the tree by strings of length m that minimizes the tree's parsimony score.

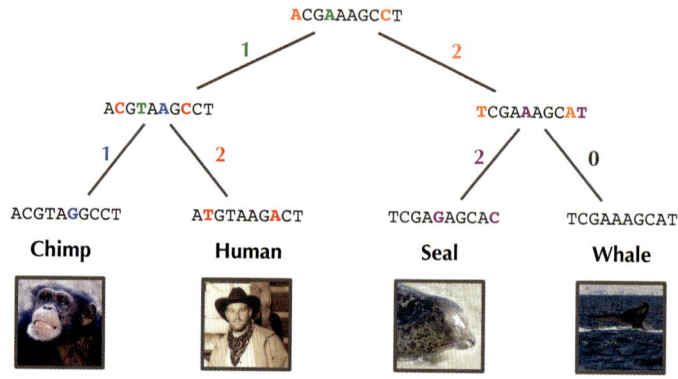

FIGURE 7.23 An evolutionary tree with parsimony score 8 whose leaves are the DNA strings in the multiple alignment from Figure 7.3. Colored letters indicate mismatches in strings connected by an edge.

Given a tree T for which every node v is labeled by a string of length m, we define labeled trees $T_1, \ldots, T_m$, where T_i has the same structure as T, and where a given node is labeled by the i-th symbol of the corresponding node in T. Since the parsimony score of T is the sum of parsimony scores of the trees $T_1, \ldots, T_m$, the Small Parsimony Problem can be solved independently for each column of the alignment. This observation lets us assume that every leaf is labeled by a single symbol rather than by a string. Thus, the weight of an edge connecting two nodes should be either 0 or 1, depending on whether these nodes are labeled by the same symbol or different symbols; given symbols i and j, we define $\delta_{i,j} = 0$ if $i = j$ and $\delta_{i,j} = 1$ if $i \neq j$.

We will describe a dynamic programming algorithm, called **SMALLPARSIMONY**, for solving the "single character" version of the Small Parsimony Problem. Recall that a rooted tree T can be viewed as a directed tree with all of its edges directed away from the root toward the leaves. Thus, every node v in T defines a subtree T_v formed by the nodes "beneath" v and consisting of all the nodes that can be reached by moving down from v (Figure 7.24).

Let k be a symbol in a given alphabet and v be a node in a tree T. Define $s_k(v)$ as the minimum parsimony score of the subtree T_v over all possible labelings of the nodes of T_v such that v is labeled by k. The initial conditions for **SMALLPARSIMONY** must assign scores to leaves. If leaf v is labeled by k, then the only character we are allowed to assign to this leaf is k. Therefore, $s_k(v) = 0$ if leaf v is labeled by symbol k, and $s_i(v) = \infty$ otherwise (Figure 7.25).

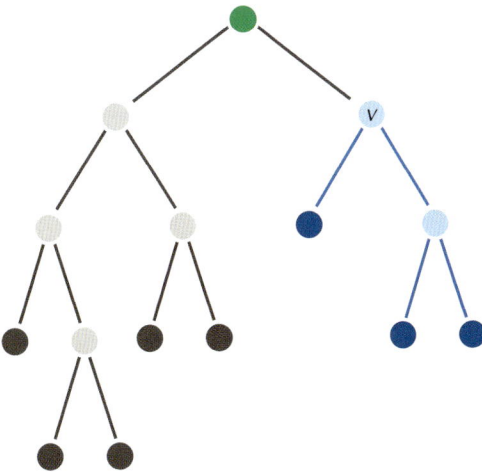

FIGURE 7.24 The (blue) subtree T_v of a node v within a larger rooted binary tree T.

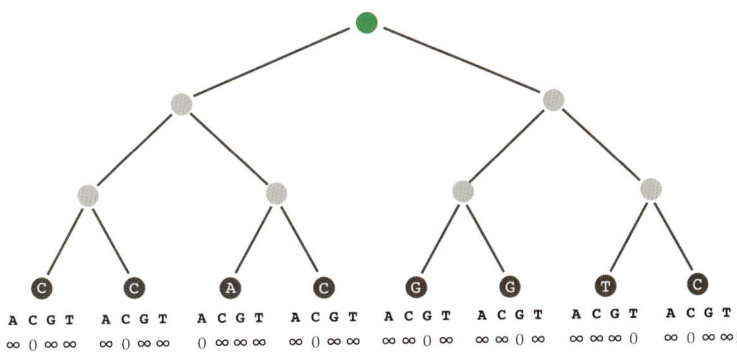

FIGURE 7.25 Initializing values $s_k(v)$ for all leaves v, represented above as an array for each leaf. We set $s_k(v)$ equal to zero if the leaf is labeled by symbol k; otherwise, we set $s_k(v)$ equal to infinity.

If v is an internal node of T, then v is connected to two "children" nodes (the nodes beneath v in T) that we arbitrarily denote as DAUGHTER(v) and SON(v). The score $s_k(v)$ can be computed as the minimum of $s_i(\text{DAUGHTER}(v)) + \delta_{i,k}$ over all possible symbols i, plus the minimum of $s_j(\text{SON}(v)) + \delta_{j,k}$ over all possible symbols j:

$$s_k(v) = \min_{\text{all symbols } i}\{s_i(\text{DAUGHTER}(v)) + \delta_{i,k}\} + \min_{\text{all symbols } j}\{s_j(\text{SON}(v)) + \delta_{j,k}\}$$

This equation allows us to compute all values $s_k(v)$ by working our way upward in T from the leaves to the root, denoted *root* (Figure 7.26). The subtree of the root is the entire tree T, and so the minimum parsimony score is given by the smallest score $s_k(root)$ over all symbols k,

$$\min_{\text{all symbols } k} s_k(root).$$

The pseudocode for **SMALLPARSIMONY** is shown below. It returns the parsimony score for a rooted binary tree T whose leaves are labeled by symbols stored in an array CHARACTER (i.e., CHARACTER(v) is the label of leaf v). At each iteration, it selects a node v and computes $s_k(v)$ for each symbol k in the alphabet. For each node v, **SMALLPARSIMONY** maintains a value TAG(v), which indicates whether the node has been processed (i.e., TAG(v) = 1 if the array $s_k(v)$ has been computed and TAG(v) = 0 otherwise). We call an internal node of T **ripe** if its tag is 0 but its children's tags are both 1. **SMALLPARSIMONY** works upward from the leaves, finding a ripe node v at which to compute $s_k(v)$ at each step.

SMALLPARSIMONY(*T*, CHARACTER)
    for each node *v* in tree *T*
        TAG(*v*) ← 0
    if *v* is a leaf
        TAG(*v*) ← 1
        for each symbol *k* in the alphabet
            if CHARACTER(*v*) = *k*
                $s_k(v) \leftarrow 0$
            else
                $s_k(v) \leftarrow \infty$
    while there exist ripe nodes in *T*
        v ← a ripe node in *T*
        TAG(*v*) ← 1
        for each symbol *k* in the alphabet
            $s_k(v) \leftarrow \min_{\text{all symbols } i} \{s_i(\text{DAUGHTER}(v)) + \delta_{i,k}\} + \min_{\text{all symbols } j} \{s_j(\text{SON}(v)) + \delta_{j,k}\}$
    return $\min_{\text{all symbols } k} s_k(v)$

STOP and Think: What is the final ripe node processed by **SMALLPARSIMONY**, regardless of the order in which we process ripe nodes?

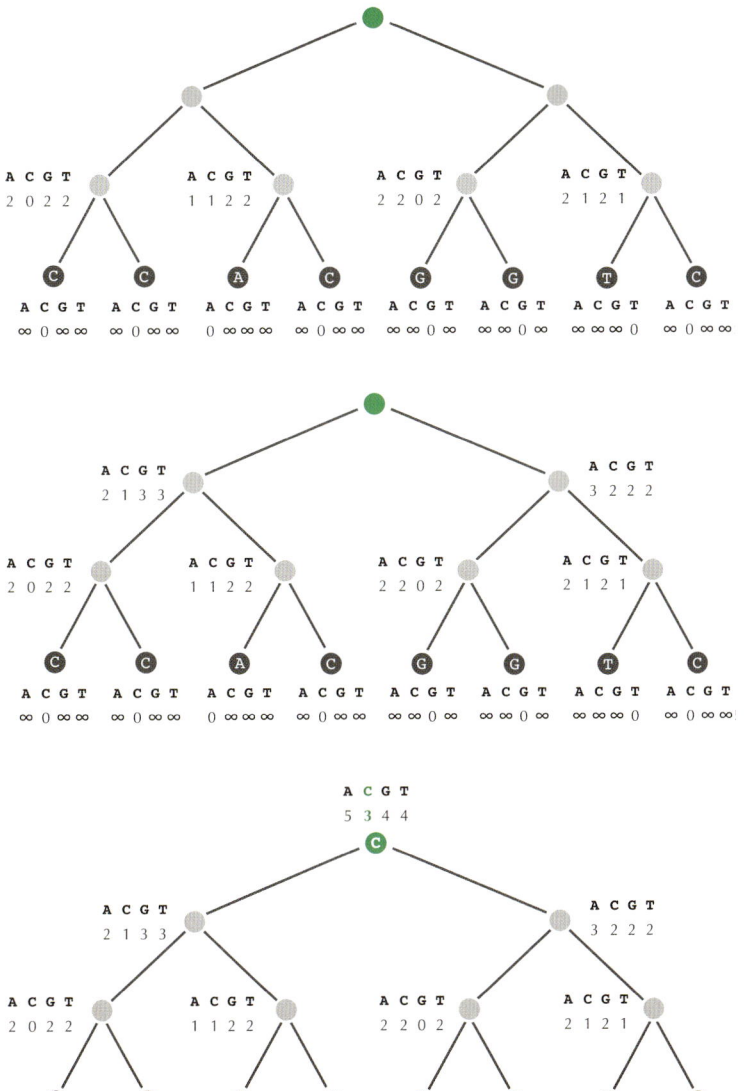

FIGURE 7.26 An illustration of **SMALLPARSIMONY** after initialization in Figure 7.25. The parsimony score is equal to the minimum score at the root, which for this tree is equal to 3. This value corresponds to symbol C, and so when we begin backtracking to assign symbols to internal nodes, we assign nucleotide C to the root.

CHAPTER 7

Once we compute the parsimony score of a tree T, we also need some way of assigning symbols to internal nodes of T. The minimum value of $s_k(root)$ in Figure 7.26 (bottom) is equal to 3, which is achieved when $k = \text{C}$. Assigning symbols to the remaining internal nodes is similar to the backtracking approach that we used for sequence alignment. Because you are already a pro at dynamic programming, we leave this task to you as an exercise.

EXERCISE BREAK: Assign nucleotides to all nodes of the tree in Figure 7.26 in order to solve the Small Parsimony Problem.

When the position of the root in the tree is unknown, we can simply assign the root to any edge that we like, apply **SMALLPARSIMONY** to the resulting tree, and then remove the root. It can be shown that this method provides a solution to the following problem.

Small Parsimony in an Unrooted Tree Problem:
Find the most parsimonious labeling of the internal nodes of an unrooted tree.

Input: An unrooted binary tree with each leaf labeled by a string of length m.
Output: A labeling of all other nodes of the tree by strings of length m that minimizes the tree's parsimony score.

EXERCISE BREAK: Estimate the runtime of the proposed algorithm for solving the Small Parsimony in an Unrooted Tree Problem.

EXERCISE BREAK: Given a multiple alignment of SARS viruses, reconstruct the most parsimonious amino acid sequence of the Spike protein in the ancestral SARS virus (the root of the tree) under the assumption that the evolutionary tree in Figure 7.17 is correct.

The Large Parsimony Problem

SMALLPARSIMONY does not help us if we do not know the evolutionary tree in advance. In this case, we must find a binary tree as well as assign ancestral strings to all internal nodes of this tree in order to minimize the parsimony score. Figure 7.27 presents the solutions of the Small Parsimony in an Unrooted Tree Problem for the three different unrooted binary trees with four leaves, where the leaves are assigned the strings from the toy multiple alignment in Figure 7.3.

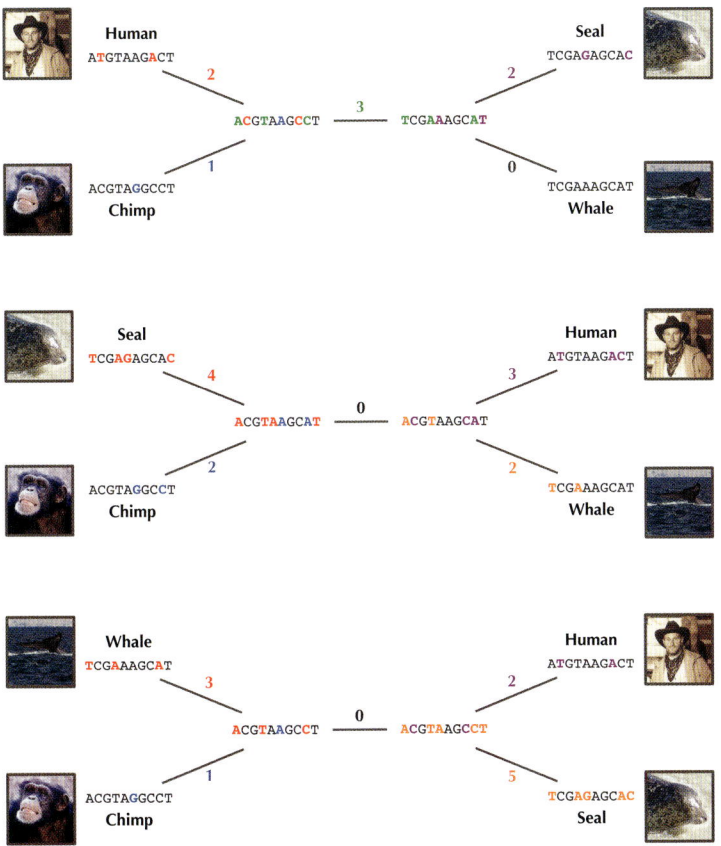

FIGURE 7.27 The three unrooted binary tree structures for the four species from Figure 7.3, with internal nodes labeled by a solution of the Small Parsimony in an Unrooted Tree Problem. The first tree solves the Large Parsimony Problem because it has smaller parsimony score (8) than the other two trees (11 each).

CHAPTER 7

Large Parsimony Problem:
Given a set of strings, find a tree — with leaves labeled by all these strings — having minimum parsimony score.

Input: A collection of strings of equal length.
Output: An unrooted binary tree T that minimizes the parsimony score among all possible unrooted binary trees with leaves labeled by these strings.

Unfortunately, the Large Parsimony problem is *NP*-Complete, in part because the number of different trees grows very quickly with respect to the number of leaves. As a workaround, we will use a greedy heuristic that explores some but not all trees. First, note that in any unrooted binary tree, the removal of an internal edge, along with the two internal nodes that this edge connects, results in four subtrees, which we will call W, X, Y, and Z (Figure 7.28). These four subtrees can be combined into a tree in three different ways, which we denote $WX|YZ$, $WY|XZ$, and $WZ|XY$. These three trees are called **nearest neighbors**; a **nearest neighbor interchange** operation replaces a tree with one of its nearest neighbors.

EXERCISE BREAK: Find the two nearest neighbors of the tree in Figure 7.15 for the internal edge connecting the parents of gorilla and human.

Like the 2-break operation in Chapter 6, a nearest neighbor interchange corresponds to replacing two edges in the tree by two new edges. For example, denote the internal edge of a nearest neighbor interchange as (a, b); denote the remaining nodes adjacent to a as w and x; and denote the remaining nodes adjacent to b as y and z. The tree on the right in Figure 7.28 is obtained from the tree on the left by removing edges (a, x) and (b, y) and replacing them with (a, y) and (b, x). The tree on the bottom in Figure 7.28 is obtained from the tree on the left by removing edges (a, x) and (b, z) and substituting them with (a, z) and (b, x).

EXERCISE BREAK: Figure 7.29 shows all possible unrooted binary trees with five leaves. Find a pair of these trees that are the "farthest apart" in that they require the maximum number of nearest neighbor interchanges to transform one tree into the other.

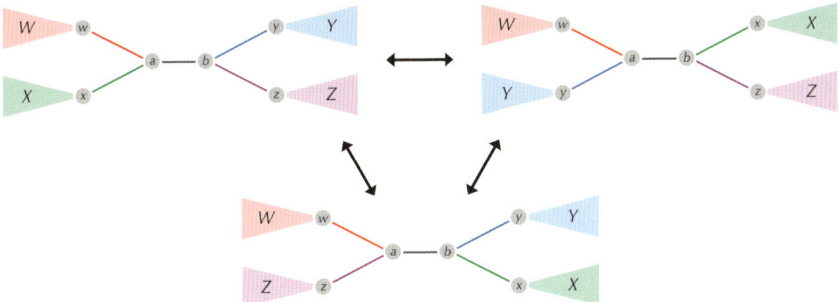

FIGURE 7.28 A nearest neighbor interchange on the internal edge (a, b), shown in black, results from rearranging the four colored subtrees W, X, Y, and Z, which are rooted at w, x, y, and z, respectively. The nearest neighbor interchange operation removes one edge connected to a and another edge connected to b, then replaces these edges with two new edges. The three possible tree structures resulting from nearest neighbor interchanges on (a, b) can be represented as $WX|YZ$ (top left), $WY|XZ$ (top right), and $WZ|XY$ (bottom).

Nearest Neighbors of a Tree Problem:

Given an edge in a binary tree, generate the tree's nearest neighbors.

    Input: An internal edge in a binary tree.
    Output: The two nearest neighbors of this tree (with respect to the given internal edge).

7H

The **nearest neighbor interchange heuristic** for the Large Parsimony Problem starts from an arbitrary unrooted binary tree. It assigns input strings to arbitrary leaves of this tree, assigns strings to the internal nodes of the tree by solving the Small Parsimony Problem in an Unrooted Tree, and then moves to a nearest neighbor that provides the best improvement in the parsimony score. At each iteration, the algorithm explores all internal edges of a tree and generates all nearest neighbor interchanges for each internal edge. For each of these nearest neighbors, the algorithm solves the Small Parsimony Problem to reconstruct the labels of the internal nodes and computes the parsimony score. If a nearest neighbor with smaller parsimony score is found, then the algorithm selects the one with smallest parsimony score (ties are broken arbitrarily) and iterates again; otherwise, the algorithm terminates. This is achieved by the following pseudocode.

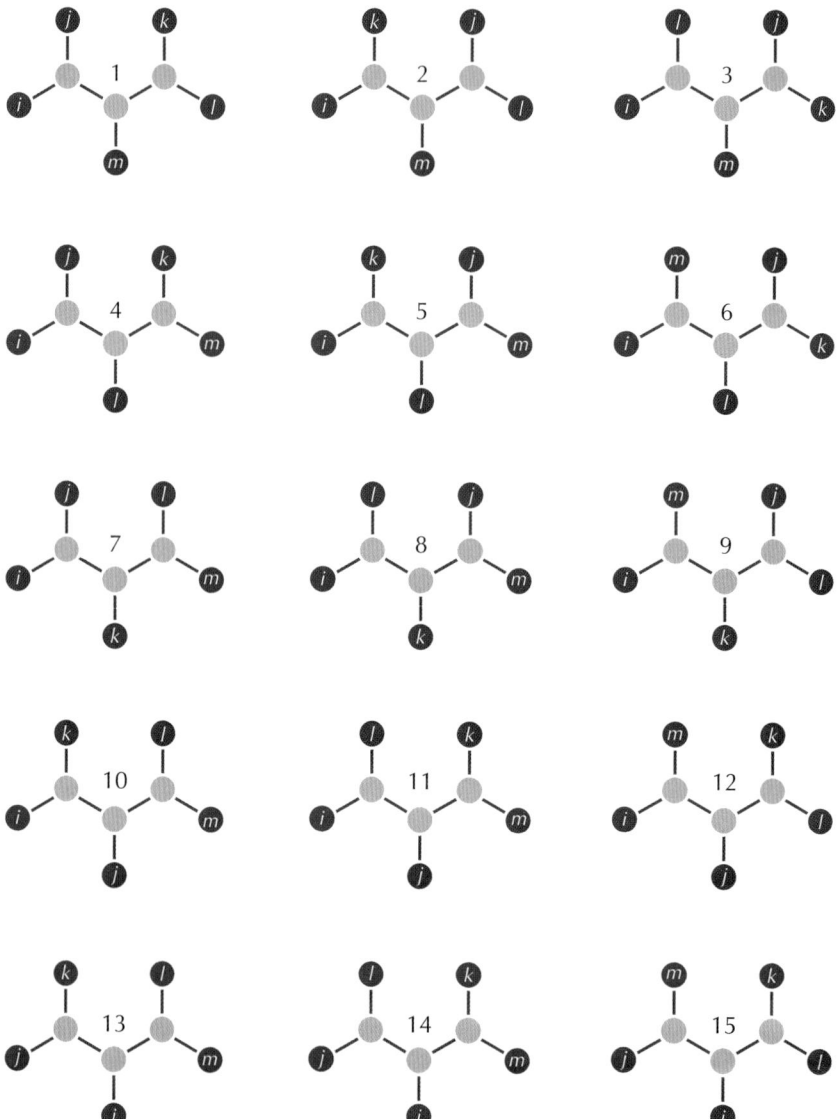

FIGURE 7.29 All fifteen unrooted binary trees with five labeled leaves. Tree 1 can be transformed into trees 4, 7, 12, and 15 by a single nearest neighbor interchange. Note that every tree has the same structure; this is not the case for trees containing more than five leaves.

NearestNeighborInterchange(*Strings*)
    score ← ∞
    generate an arbitrary unrooted binary tree *Tree* with |*Strings*| leaves
    label the leaves of *Tree* by arbitrary strings from *Strings*
    solve the Small Parsimony in an Unrooted Tree Problem for *Tree*
    label the internal nodes of *Tree* according to a most parsimonious labeling
    newScore ← the parsimony score of *Tree*
    newTree ← *Tree*
    while *newScore* < *score*
        score ← *newScore*
        Tree ← *newTree*
        for each internal edge *e* in *Tree*
            for each nearest neighbor *NeighborTree* of *Tree* with respect to the edge *e*
                solve the Small Parsimony in an Unrooted Tree Problem for *NeighborTree*
                neighborScore ← the minimum parsimony score of *NeighborTree*
                if *neighborScore* < *newScore*
                      newScore ← *neighborScore*
                      newTree ← *NeighborTree*
    return *newTree*

EXERCISE BREAK: Chapter 8 describes how the alcohol dehydrogenase (Adh) gene helps yeast produce alcohol. To study the evolution of this gene, biologists constructed a multiple alignment of the Adh genes from various yeast species. Use this alignment to reconstruct the evolutionary tree of various yeast species and an ancient yeast Adh ancestor gene.

We have now encountered a number of algorithms for constructing evolutionary trees, but this does not mean that we can easily resolve various evolutionary controversies. For example, the identity of the chimpanzee's closest relative remained undecided until the mid-1990s (Figure 7.30), and the question of whether mice are closer to humans than to dogs is still the subject of debate (Figure 7.31).

CHAPTER 7

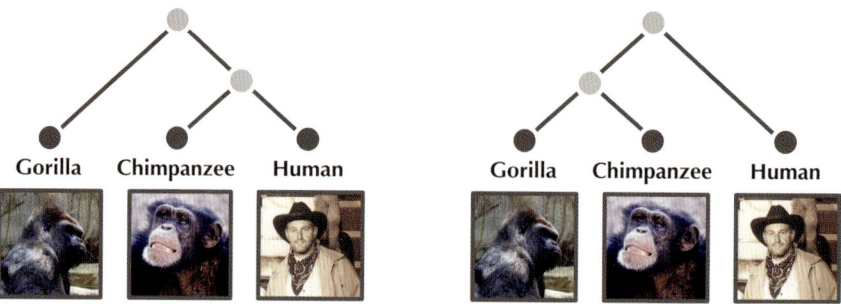

FIGURE 7.30 (Left) Analysis of beta-globin genes in human, chimpanzee, and gorilla suggests a human-chimpanzee split. (Right) Analysis of dopamine D4 receptor gene suggests a gorilla-chimpanzee split.

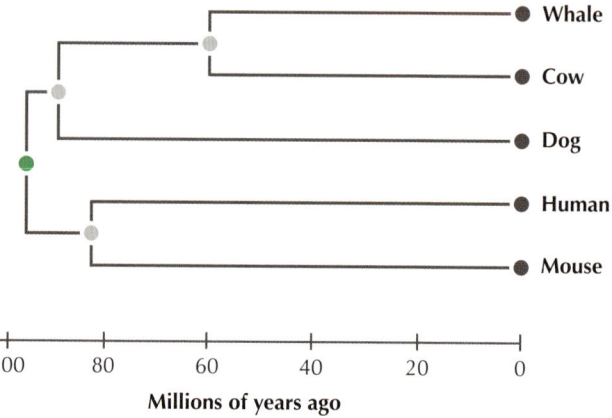

FIGURE 7.31 At the start of the 21st Century, biologists believed that dogs are evolutionarily closer to humans than mice are. However, recent studies suggest otherwise, as shown in the above phylogeny.

Epilogue: Evolutionary Trees Fight Crime

Janice Trahan met Dr. Richard Schmidt in 1982 when she began working as a nurse in Lafayette, Louisiana. Both Janice and Richard were married with children, but they fell in love. Janice soon divorced her husband, but although Richard promised that he would divorce his wife, he never did. After twelve years, she got tired of waiting, and broke off the relationship. Two weeks later, she awoke in the middle of the night to see Richard standing over her, a syringe in his hand.

Although Janice had broken up with Richard, she was not surprised to see him. She had even left her door unlocked for him because he had been giving her vitamin B-12 injections for her chronic fatigue. Since Janice had B-12 injections before, she knew what to expect. This time, however, she experienced scorching pain as Richard squeezed the syringe.

A few months later, Janice tested positive for HIV, and she accused Richard of infecting her via an injection. The police detective in Lafayette had never heard anything as bizarre as this revenge story — a syringe tainted with HIV had never been used as a murder weapon. At first, he suspected that Janice had fabricated the story to tarnish the reputation of her former lover. Nevertheless, he started an investigation by collecting samples from several HIV patients in Lafayette.

After investigating hospital records, the detective found that Richard had taken blood from Donald McClelland, an HIV patient, on the same day that he had injected Janice. Now the forensic challenge was to determine whether the HIV strain taken from McClelland was similar to Janice's strain. After HIV DNA was collected from Janice, McClelland, and many other unrelated HIV-infected patients from Lafayette, scientists constructed the evolutionary tree of these HIV viruses and found that the viruses sampled from Janice and McClelland formed a subtree of this tree (Figure 7.32).

The case "State of Louisiana vs. Richard Schmidt" went to trial in 1998. The prominent evolutionary biologist David Hillis presented the evolutionary tree as evidence of the crime, demonstrating that Janice's HIV sequence had been derived (with some small variations) from McClelland's HIV sequence. Richard Schmidt was then sentenced to 50 years in prison for attempted murder.

STOP and Think: If you had been Richard Schmidt's attorney, how would you have argued for his innocence?

CHALLENGE PROBLEM: Given HIV sequences from AIDS patients in Lafayette, construct the evolutionary tree for other HIV proteins. Does each tree support conviction of Dr. Schmidt? Reconstruct the ancestral HIV sequences at the internal nodes of the resulting trees.

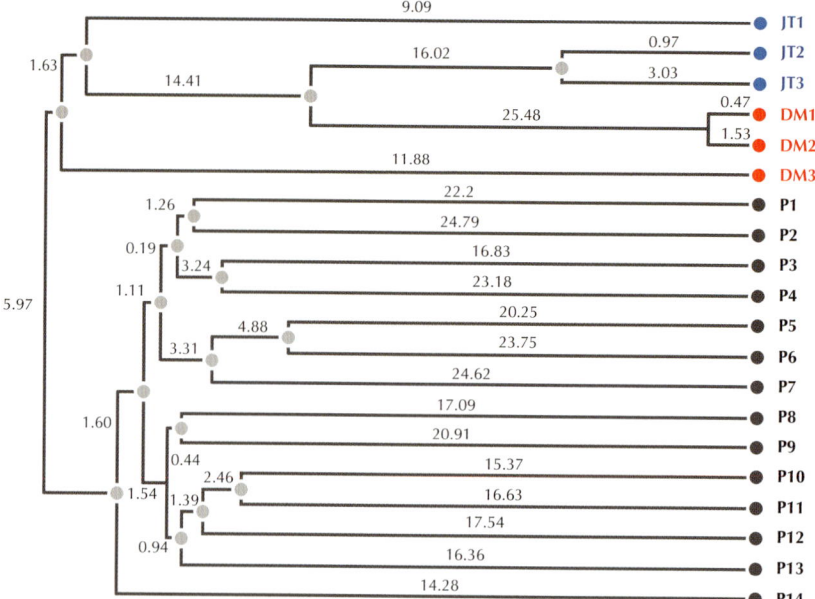

FIGURE 7.32 An evolutionary tree of HIV viruses taken from various patients in Lafayette. Samples from the victim, Janice Trahan (blue leaves JT1, JT2, and JT3), and Richard Schmidt's patient, Donald McClelland (red leaves DM1, DM2, DM3), are clustered together and are rather different than sequences from other patients from Lafayette (labeled P1 to P14).

Detours

When did HIV jump from primates to humans?

Scientists became aware that HIV causes AIDS in the 1980s, at a time when the virus was still rare. They immediately started looking for earlier cases of HIV infection in various medical records and found HIV in a blood sample taken from a Congolese patient in 1959. Genetic studies then revealed that HIV is closely related to Simian Immunodeficiency Virus (SIV), which infects primates, but it remained unclear how and when SIV entered the human population and evolved into HIV. The most popular current hypothesis is that SIV evolved into HIV when hunters killed monkeys to sell their meat and were exposed to the animals' blood. The virus circulating in the blood had further entered cuts in the hunters' skin, mutated, and later adapted to humans.

By finding regions in the viral genome that mutate at a roughly constant rate over time, researchers can infer a timeline of HIV evolution. Using this "molecular clock", biologists estimated the time points when the various subtypes of SIV jumped from primates to human: for HIV groups A, B, M, and O, these transitions have been estimated at 1940, 1945, 1908, and 1920, respectively (the timing of the jump for group N is currently unknown). By sequencing fecal samples of wild primates, biologists have even been able to locate populations of chimpanzees and sooty mangabeys whose SIVs are direct ancestors of HIV groups.

Searching for a tree fitting a distance matrix

Every 3×3 matrix D is additive. To see why, note that there is only one tree with three leaves. We will denote the leaves of this tree as 1, 2, and 3, and its internal node as c. As illustrated in Figure 7.33, the lengths of edges in this tree must satisfy the following three equations:

$$d_{1,c} + d_{2,c} = D_{1,2} \qquad d_{1,c} + d_{3,c} = D_{1,3} \qquad d_{2,c} + d_{3,c} = D_{2,3}$$

Solving this system of equations yields the following formulas for the lengths of edges in terms of values of the matrix D:

$$d_{1,c} = \frac{D_{1,2} + D_{1,3} - D_{2,3}}{2} \qquad d_{2,c} = \frac{D_{2,1} + D_{2,3} - D_{1,3}}{2} \qquad d_{3,c} = \frac{D_{3,1} + D_{3,2} - D_{1,2}}{2}$$

Figure 7.34 illustrates an attempt to fit the distance matrix in Figure 7.9 (right) to all possible unrooted trees with four leaves. Each such tree leads to a system of six linear equations (in either four or five variables) that does not have a solution. Thus, this distance matrix must be non-additive.

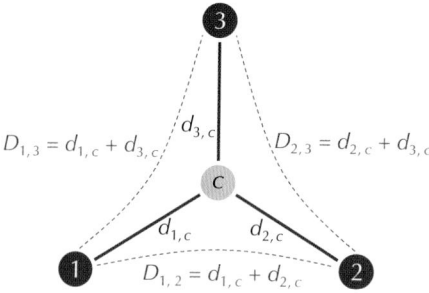

FIGURE 7.33 The tree with three leaves 1, 2, and 3 in addition to an internal node c. The distances between leaves ($D_{1,2}$, $D_{1,3}$, and $D_{2,3}$) uniquely define the lengths of edges ($d_{1,c}$, $d_{2,c}$, and $d_{3,c}$).

The reason we failed to fit a tree to the 4×4 distance matrix in Figure 7.34 is partly because the number of equations was larger than the number of variables ($d_{i,j}$) for each tree (a fact that will also hold for larger values of n). If the number of equations in a linear system of equations is smaller than or equal to the number of variables in the system, then a solution usually exists, whereas if the number of equations exceeds the number of variables, there is usually no solution. However, there are exceptions in both directions. Consult an introductory linear algebra text for more details.

The four point condition

The **four point condition** gives an alternative way to determine if a matrix is additive. Consider the tree containing only four leaves in Figure 7.35. For this tree, observe that

$$d_{i,j} + d_{k,l} \leq d_{i,k} + d_{j,l} = d_{i,l} + d_{j,k}$$

because the first sum is the sum of lengths of all edges in the tree *minus* the length of the internal edge, while the last two sums are equal to the sum of lengths of all edges in the tree *plus* the length of the internal edge.

Indeed, for any quartet of leaves (i, j, k, l) in an arbitrary tree, if we compute the three sums

$$d_{i,j} + d_{k,l} \quad d_{i,k} + d_{j,l} \quad d_{i,l} + d_{j,k}$$

then we will find that two of the sums are equal and that the third sum is less than or equal to the other two sums. In terms of an $n \times n$ distance matrix, we say that a quartet of indices (i, j, k, l) satisfies the four point condition if two of the following sums are equal, and the third sum is less than or equal to the other two sums:

WHICH ANIMAL GAVE US SARS?

$$D_{i,j} + D_{k,l} \quad D_{i,k} + D_{j,l} \quad D_{i,l} + D_{j,k}$$

Four Point Theorem: *A distance matrix is additive if and only if the four point condition holds for every quartet (i, j, k, l) of indices of this matrix.*

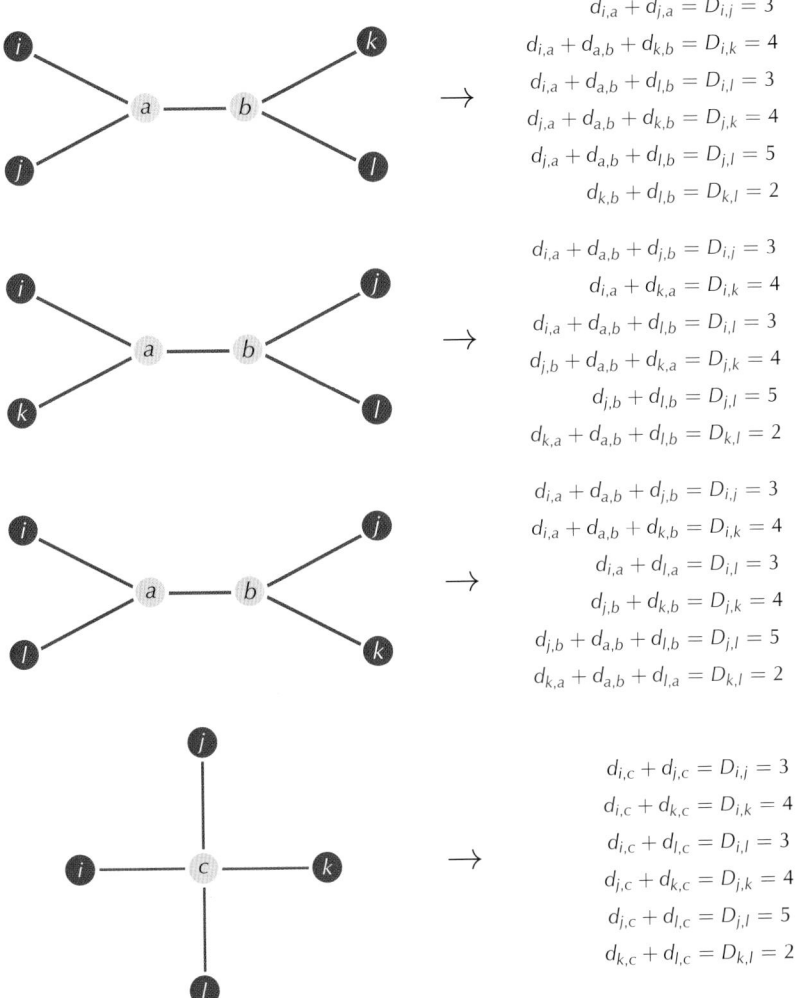

FIGURE 7.34 (Left) All four trees with four leaves. (Right) Each attempt to fit the distance matrix in Figure 7.9 to a tree results in a system of six linear equations. Because none of these systems has a solution, this matrix must be non-additive.

CHAPTER 7

EXERCISE BREAK: Prove the Four Point Theorem.

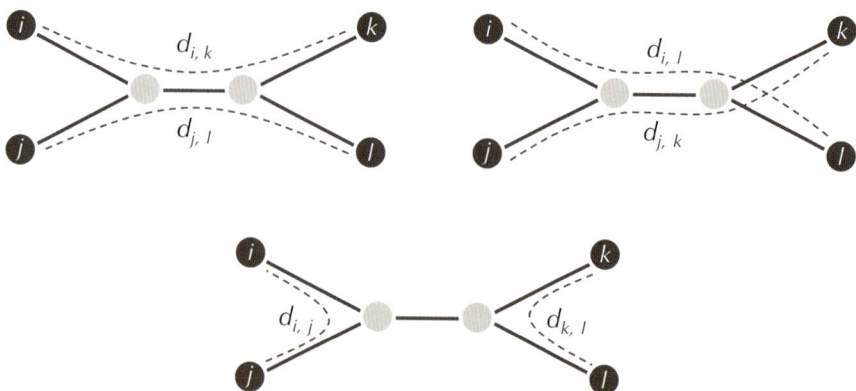

FIGURE 7.35 Three pairs of paths through a tree with four leaves. The paths in the top left and top right traverse the same edges, and so $d_{i,k} + d_{j,l} = d_{i,l} + d_{j,k}$. Furthermore, $d_{i,j} + d_{k,l}$ must be less than or equal to these two sums, since it does not traverse the internal edge of the tree, as shown in the tree on the bottom.

The Four Point Theorem provides us with an alternative way of determining whether a given distance matrix is additive, since we can simply check whether the four point condition holds for each quartet of indices of the matrix.

EXERCISE BREAK: Compare the running time of this proposed method with that of a variant of **ADDITIVEPHYLOGENY** deciding whether a given distance matrix is additive.

EXERCISE BREAK: Find a quartet of indices from the distance matrix in Figure 7.9 (right) that violates the four points condition.

Did bats give us SARS?

During the search for the animal reservoir of the SARS virus, biologists discovered infected palm civets at a live animal market in China. Meat from these animals is often added to "dragon-tiger-phoenix soup", an expensive Cantonese dish. The discovery did not greatly change the infected civets' fate: instead of ending up in soup, they were made into SARS scapegoats and slaughtered anyway.

Yet when further searches failed to identify more SARS-infected civets, biologists started to wonder whether palm civets really were the original source of SARS. In 2005, they discovered a SARS-like virus in Chinese horseshoe bats (Figure 7.36). The bats turned out to be SARS-CoV carriers, but they can probably only pass the virus to humans through intermediate hosts. Since bat meat is considered a delicacy and is also used in traditional Chinese medicine, bats had plenty of chances to come in close contact with civets at overcrowded live animal markets.

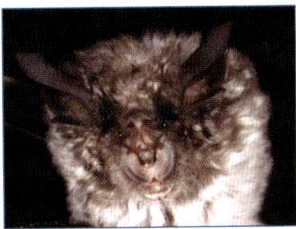

FIGURE 7.36 The horseshoe bat.

When biologists constructed the evolutionary tree of coronaviruses from bats, civets, and humans (Figure 7.37), they found that both the civet and human variants of SARS-CoV are nested within a bat virus phylogeny.

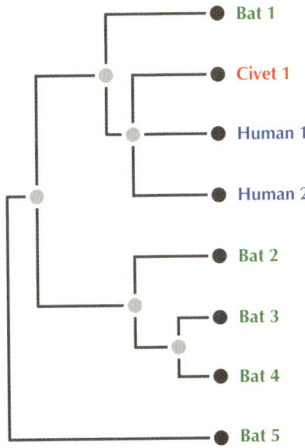

FIGURE 7.37 An evolutionary tree of coronaviruses from bats, civets, and humans.

CHAPTER 7

Even before sequencing SARS-CoV, biologists knew of other human coronaviruses but considered them harmless. However, a new deadly SARS-like coronavirus emerged in 2012 in Saudi Arabia, This virus (causing **Middle East Respiratory Syndrome (MERS)**) generated international headlines as it quickly spread to other countries.

Although researchers initially believed that camels were to blame for MERS — many Saudis consume unpasteurized camel milk — most afflicted patients had not come in contact with camels. Yet when researchers tested a coronavirus taken from a bat found just a few miles from one of the first MERS patient's homes, they found this virus to be a nearly perfect match with the patient's virus sample.

Why does the neighbor-joining algorithm find neighboring leaves?

We stated in the main text that if the distance matrix D is additive, then there exists a unique simple tree $\text{TREE}(D)$ fitting this matrix. This is not quite true, since if a simple tree fits D and has an internal edge of weight zero, then we can easily remove this edge by "gluing" together the nodes that it connects (Figure 7.38). Thus, we will make a further assumption that not only is $\text{TREE}(D)$ simple, but that it contains no internal edges of length zero.

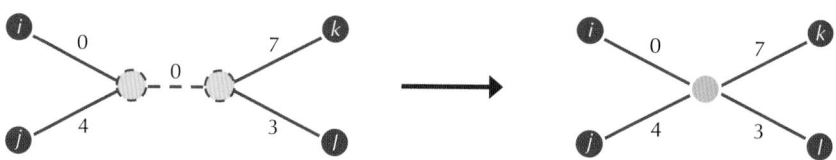

FIGURE 7.38 Gluing the nodes at the endpoints of an internal edge having zero length (shown by the dashed edge).

Now, we can rewrite the neighbor-joining matrix formula as follows:

$$D^*_{i,j} = (n-2) \cdot D_{i,j} - \text{TOTALDISTANCE}_D(i) - \text{TOTALDISTANCE}_D(j)$$
$$= (n-2) \cdot D_{i,j} - \sum_{1 \leq k \leq n} D_{i,k} - \sum_{1 \leq k \leq n} D_{j,k}.$$

Each element in this formula further breaks down as a sum of edge weights in $\text{TREE}(D)$.

For example, for the first tree in Figure 7.34,

$$
\begin{aligned}
D^*_{1,2} &= 2 \cdot d_{1,2} - (d_{1,3} + d_{1,4} + d_{1,2}) - (d_{2,3} + d_{2,4} + d_{1,2}) \\
&= 2 \cdot (d_{1,a} + d_{a,2}) - ((d_{1,a} + d_{a,b} + d_{b,3}) + (d_{1,a} + d_{a,b} + d_{b,4}) + \\
&\quad (d_{1,a} + d_{a,2})) - ((d_{2,a} + d_{a,b} + d_{b,3}) + (d_{2,a} + d_{a,b} + d_{b,4}) + (d_{1,a} + d_{a,2})) \\
&= -2 \cdot d_{1,a} - 2 \cdot d_{2,a} - 2 \cdot d_{b,3} - 2 \cdot d_{b,4} - 4 \cdot d_{a,b},
\end{aligned}
$$

and

$$
\begin{aligned}
D^*_{1,3} &= 2 \cdot d_{1,3} - (d_{1,2} + d_{1,4} + d_{1,3}) - (d_{3,4} + d_{3,2} + d_{3,1}) \\
&= 2 \cdot (d_{1,a} + d_{a,b} + d_{b,3}) - ((d_{1,a} + d_{a,2}) + (d_{1,a} + d_{a,b} + d_{b,4}) + \\
&\quad (d_{1,a} + d_{a,b} + d_{b,3})) - ((d_{3,b} + d_{b,4}) + (d_{3,b} + d_{b,a} + d_{b,2}) + (d_{3,b} + d_{b,a} + d_{a,1})) \\
&= -2 \cdot d_{1,a} - 2 \cdot d_{2,a} - 2 \cdot d_{b,3} - 2 \cdot d_{b,4} - 2 \cdot d_{a,b}.
\end{aligned}
$$

Note that the expressions for $D^*_{1,2}$ and $D^*_{1,3}$ are nearly identical, with only the coefficients of $d_{a,b}$ differing (highlighted in red above). Since $D^*_{1,2} - D^*_{1,3} = -2 \cdot d_{a,b} < 0$, the neighbor-joining algorithm will prefer the smaller $D^*_{1,2}$ over $D^*_{1,3}$.

Given an edge e in TREE(D), the **multiplicity** of this edge in $D^*_{i,j}$ is the coefficient of d_e in $D^*_{i,j}$, denoted MULTIPLICITY$_{i,j}(e)$,. For example, for the edge $e = (a,b)$ in the first tree in Figure 7.34, MULTIPLICITY$_{1,2}(e) = -4$, and MULTIPLICITY$_{1,3}(e) = -2$. The following result shows that all limbs in TREE(D) have the same multiplicity.

Lemma. *For an additive distance matrix D and any pair of leaves i and j in* TREE(D), MULTIPLICITY$_{i,j}(e)$ *is equal to* -2 *for any limb e in* TREE(D).

Proof. If a limb e is not the limb of leaf i or leaf j, then it is counted zero times in $(n-2) \cdot D_{i,j}$, once in TOTALDISTANCE$_D(i)$ and once in TOTALDISTANCE$_D(j)$, making its multiplicity -2. On the other hand, if e is the limb of i or j (say, i), then it is counted $n-2$ times in $(n-2) \cdot D_{i,j}$, $n-1$ times in TOTALDISTANCE$_D(i)$, and once in TOTALDISTANCE$_D(j)$. Therefore, its multiplicity is $n - 2 - (n-1) - 1 = -2$. □

This lemma implies that regardless of which pair of leaves i and j we choose, the limbs of TREE(D) will all have the same contribution to the computation of $D^*_{i,j}$. As a result, only the multiplicities of internal edges of TREE(D) differentiate values of the matrix D^*. Can we determine these multiplicities?

> **EXERCISE BREAK:** Prove that for any i and j, and for any internal edge e in TREE(D), MULTIPLICITY$_{i,j}(e) \leq -2$.

Not only does the multiplicity of each internal edge not exceed -2, but we also have a condition determining when an internal edge will have multiplicity equal to -2. To derive this condition, first note that the removal of an internal edge e disconnects any tree into two subtrees. If e lies on the (unique) path connecting leaves i and j in TREE(D), denoted PATH(i, j), then i and j belong to different subtrees, denoted T_i and T_j, respectively; otherwise, i and j belong to the same subtree (Figure 7.39). In the latter case, we denote the number of leaves in the subtree that *does not* include leaves i and j as LEAVES$_{i,j}(e)$.

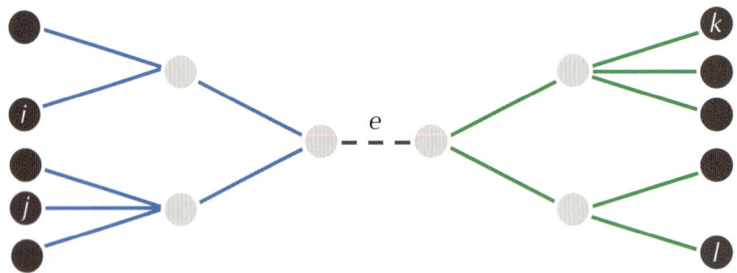

FIGURE 7.39 If an internal edge e lies on the unique path connecting two leaves in the original tree (such as j and k), then these leaves are separated into different subtrees (shown as blue and green) after removing e. If e does not lie on the unique path connecting two leaves (such as k and l), then these leaves belong to the same subtree after removing e.

Edge Multiplicity Theorem: *Given an additive matrix D, the multiplicity of an internal edge e is equal to -2 if e lies on* PATH(i, j) *in* TREE(D) *and is equal to* $-2 \cdot$ LEAVES$_{i,j}(e)$ *otherwise.*

Proof. If an internal edge e lies on PATH(i, j) in TREE(D), then e has a coefficient of $n - 2$ in $(n - 2) \cdot D_{i,j}$ term in $D^*_{i,j}$. To compute MULTIPLICITY$_{i,j}(e)$, consider the subtrees T_i and T_j formed by the removal of e. For each leaf k in T_i, PATH(j, k) passes through e, thus contributing 1 to the coefficient of e in TOTALDISTANCE$_D(j)$, but PATH(i, k) does not pass through e, thus contributing 0 to the coefficient of e in TOTALDISTANCE$_D(i)$).

Likewise, for each leaf k in T_j, PATH(i, k) passes through e, contributing 1 to the coefficient of e in TOTALDISTANCE$_D(i)$, but PATH(j, k) does not pass through e, contributing 0 to the coefficient of e in TOTALDISTANCE$_D(i)$). As a result, every leaf k will contribute 1 to the coefficient of e either in TOTALDISTANCE$_D(i)$ or in TOTALDISTANCE$_D(j)$. The coefficient of e in TOTALDISTANCE$_D(i)$ + TOTALDISTANCE$_D(j)$ is n, which means that

$$\text{MULTIPLICITY}_{i,j}(e) = (n-2) - n = -2.$$

On the other hand, if e does not lie on $\text{PATH}(i,j)$, then e has a coefficient of 0 in $(n-2) \cdot D_{i,j}$. And if k is a leaf in the subtree not containing i and j, then to reach k, we must pass through e. Thus, the coefficient of e in each of $\text{TOTALDISTANCE}_D(i)$ and $\text{TOTALDISTANCE}_D(j)$ is equal to $\text{LEAVES}_{i,j}(e)$. It follows that

$$\text{MULTIPLICITY}(e) = 0 - 2 \cdot \text{LEAVES}_{i,j}(e) = -2 \cdot \text{LEAVES}_{i,j}(e).$$

□

We can interpret the Edge Multiplicity Theorem as stating that internal edges on the path $\text{PATH}(i,j)$ have large multiplicities (-2) and other internal edges have small multiplicities (less than -2). Thus, if we are trying to minimize $D_{i,j}^*$ (among all possible choices of i and j), then we should look for a pair of leaves (i, j) having few internal edges on $\text{PATH}(i,j)$. Neighbors have no internal edges connecting them, which makes them attractive candidates. The following exercise will get us part of the way toward proving that neighbors indeed minimize $D_{i,j}^*$.

EXERCISE BREAK: Show that if leaves i and j are neighbors in $\text{TREE}(D)$, and leaf k is not a neighbor of i, then $D_{i,j}^* < D_{i,k}^*$.

Neighbor-Joining Theorem: *Given an additive distance matrix D, a minimum element $D_{i,j}^*$ of the neighbor-joining matrix D^* corresponds to neighboring leaves i and j in $\text{TREE}(D)$.*

Proof. Assume that $D_{i,j}^*$ is a minimum element of D^* but that i and j are not neighbors in $\text{TREE}(D)$. We aim to reach a contradiction by finding a pair of neighbors k and l such that $D_{k,l}^* < D_{i,j}^*$. By the preceding exercise, neither i nor j can have a neighbor if $D_{i,j}^*$ is a minimum element of D^*. Thus, i and j are the only leaves connected to $\text{PARENT}(i)$ and $\text{PARENT}(j)$, respectively. Because $\text{TREE}(D)$ is simple, $\text{PARENT}(i)$ and $\text{PARENT}(j)$ have degree at least equal to 3, meaning that each of these nodes is connected to at least two other nodes of $\text{TREE}(D)$, one of which lies on $\text{PATH}(i,j)$. The other node is part of its own subtree; we call these subtrees T_1 and T_2 (Figure 7.40).

Without loss of generality, we will assume that the number of leaves in T_1 does not exceed the number of leaves in T_2. Because i and j are not in T_1 or T_2, T_1 must therefore contain fewer than $n/2$ leaves, and the rest of $\text{TREE}(D)$ must contain more than $n/2$ leaves (recall that n refers to the total number of leaves in $\text{TREE}(D)$). Because i

CHAPTER 7

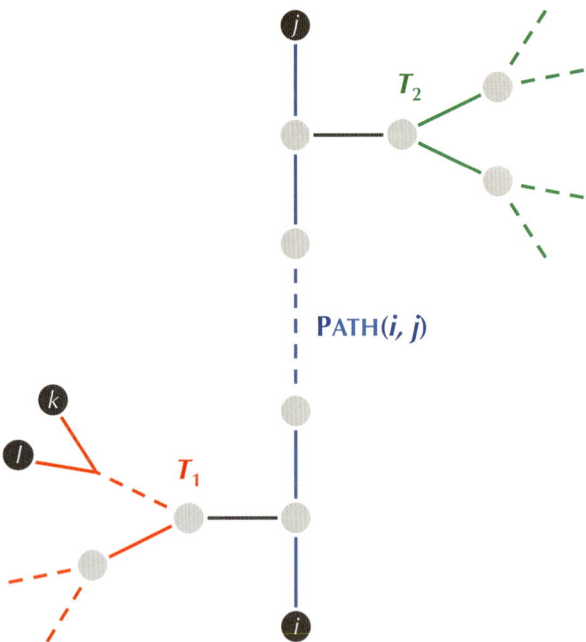

FIGURE 7.40 Leaves i and j are not neighbors. The unique path connecting them in TREE(D), PATH(i, j), is shown in blue. Because TREE(D) is simple, PARENT(i) and PARENT(j) must be connected to at least two other internal nodes, thus forming subtrees T_1 and T_2 (shown in red and green).

has no neighbors, T_1 must have at least two leaves, which implies that T_1 has a *pair* of neighbors, which we denote as (k, l). We will show that $D^*_{k,l} < D^*_{i,j}$.

Consider an internal edge e of TREE(D). We will first show that $D^*_{k,l} \leq D^*_{i,j}$ by showing that the multiplicity of e in $D^*_{k,l}$ does not exceed the multiplicity of e in $D^*_{i,j}$. There are three possibilities.

- If e lies on PATH(i, j), then by the Edge Multiplicity Theorem, MULTIPLICITY$_{i,j}(e) = -2$, and the result follows.

- If e lies on PATH(i, k), then the removal of e breaks TREE(D) into two subtrees, one containing k and l (with number of leaves equal to LEAVES$_{i,j}(e) < n/2$), and the other containing i and j (with number of leaves equal to LEAVES$_{k,l}(e) > n/2$). Thus, by the Edge Multiplicity Theorem, MULTIPLICITY$_{k,l}(e) = -2 \cdot$ LEAVES$_{k,l}(e) <$ MULTIPLICITY$_{i,j}(e) = -2 \cdot$ LEAVES$_{i,j}(e)$.

- If e lies on neither PATH(i,j) nor PATH(i,k), then the subtree not containing i and j when e is removed is the same as the subtree not containing k and l. As a result, we have that LEAVES$_{i,j}(e)$ = LEAVES$_{k,l}(e)$, which in turn implies that MULTIPLICITY$_{i,j}(e)$ = MULTIPLICITY$_{k,l}(e)$ (the Edge Multiplicity Theorem).

To prove that $D^*_{k,l}$ is in fact less than $D^*_{i,j}$, note that PATH(i,k) must contain an internal edge e because i and k are not neighbors. By the middle case above, we know that MULTIPLICITY$_{k,l}(e)$ < MULTIPLICITY$_{i,j}(e)$. Our assumption that no internal edges of TREE(D) have length zero means that d_e must be positive, which yields the result. □

Computing limb lengths in the neighbor-joining algorithm

In the main text, we were attempting to assign limb lengths to leaves of a tree constructed from an arbitrary distance matrix. We set the limb length of i equal to $\frac{1}{2}(D_{i,j} + \Delta_{i,j})$ and the limb length of j equal to $\frac{1}{2}(D_{i,j} - \Delta_{i,j})$, where

$$\Delta_{i,j} = \frac{\text{TOTALDISTANCE}_D(i) - \text{TOTALDISTANCE}_D(j)}{n-2}.$$

Where do these formulas come from?

Assume for a moment that D is an additive matrix, and select some leaf k that is not equal to i or j. If m is the parent of i and j, then we already have the limb length formula

$$\text{LIMBLENGTH}(i) = \frac{D_{i,j} + D_{i,k} - D_{j,k}}{2}.$$

As a result, it might seem like we should use this formula in the neighbor-joining algorithm for an arbitrary distance matrix D. However, if D is non-additive, then the expression $(D_{i,j} + D_{i,k} - D_{j,k})/2$ will vary depending on how we select k. So we need a formula that still computes LIMBLENGTH(i) when D is additive but that also gives us a single value when D is non-additive. To this end, we can compute the average of the above formula over all choices of $n - 2$ leaves k:

$$\frac{1}{n-2} \cdot \sum_{\text{all leaves } k \neq i,j} \frac{D_{i,j} + D_{i,k} - D_{j,k}}{2}$$

If D is additive, then the sum above contains $n - 2$ terms, all of which are equal to LIMBLENGTH(i). Furthermore, if D is non-additive, then this formula provides us with an estimate of the limb length. Note also that the sum above has $n - 2$ occurrences of

$D_{i,j}$. If we separate these out, then we obtain

$$
\begin{aligned}
\text{LimbLength}(i) &= \frac{D_{i,j}}{2} + \frac{1}{n-2} \cdot \sum_{\text{all leaves } k \neq i,j} \frac{D_{i,k} - D_{j,k}}{2} \\
&= \frac{D_{i,j}}{2} + \frac{1}{n-2} \cdot \left(\sum_{\text{all leaves } k \neq i,j} \frac{D_{i,k}}{2} - \sum_{\text{all leaves } k \neq i,j} \frac{D_{j,k}}{2} \right) \\
&= \frac{1}{2} \cdot \left(D_{i,j} + \frac{1}{n-2} \cdot \left(\sum_{\text{all leaves } k \neq i,j} D_{i,k} - \sum_{\text{all leaves } k \neq j} D_{j,k} \right) \right) \\
&= \frac{1}{2} \cdot \left(D_{i,j} + \frac{\text{TotalDistance}_D(i) - \text{TotalDistance}_D(j)}{n-2} \right) \\
&= \frac{1}{2} \cdot \left(D_{i,j} + \Delta_{i,j} \right),
\end{aligned}
$$

which is the formula in the main text that we used for computing LimbLength(i).

Giant panda: bear or raccoon?

For many years, biologists could not agree on whether the giant panda should be classified as a bear or as a raccoon. Although giant pandas look like bears, they have features that are unusual for bears and typical of raccoons: they do not hibernate in the winter, and their male genitalia are tiny and backward-pointing. As a result, Edwin Colbert wrote in 1938:

> *So the quest has stood for many years with the bear proponents and the raccoon adherents and the middle-of-the-road group advancing their several arguments with the clearest of logic, while in the meantime the giant panda lives serenely in the mountains of Szechuan with never a thought about the zoological controversies he is causing by just being himself.*

Whereas analysis of anatomical and behavioral characters only led to unsettled debates, the analysis of genetic characters by Stephen O'Brien in 1985 demonstrated that giant pandas are indeed more closely related to bears than raccoons (Figure 7.41).

Where did humans come from?

In 1987, Rebecca Cann, Mark Stoneking and Allan Wilson constructed an evolutionary tree of **mitochondrial DNA (mtDNA)** of 133 people representing African, Asian, Australian, Caucasian, and New Guinean ethnic groups. This tree led to the **Out of Africa**

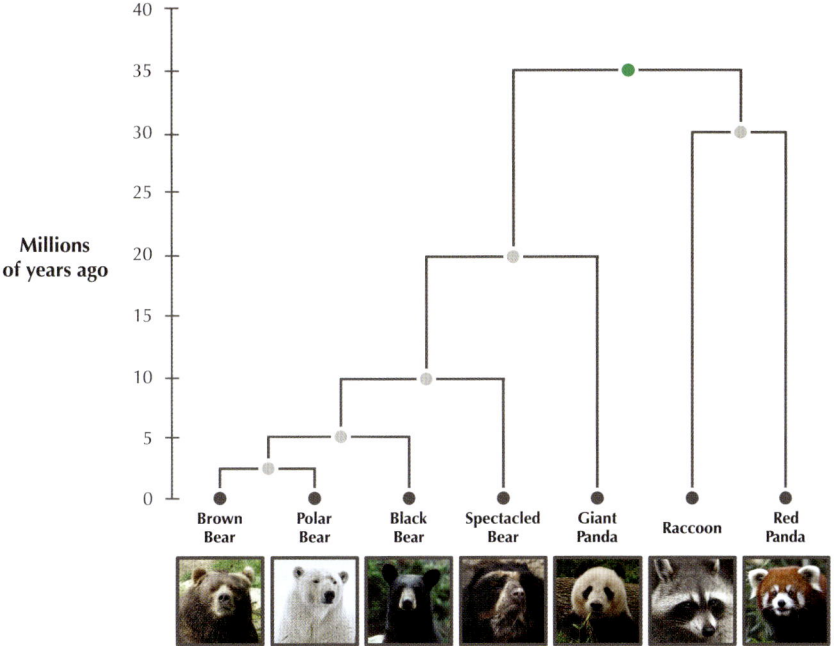

FIGURE 7.41 An evolutionary tree of bears and raccoons.

hypothesis, which claims that humans have a common ancestor who lived in Africa. This study turned the question of human origins into an algorithmic problem.

The mtDNA evolutionary tree showed a trunk splitting into two major branches (Figure 7.42). One branch, containing the bottom five individuals in Figure 7.42, consisted only of Africans, whereas the other branch included some modern Africans as well as all people belonging to other ethnic groups.

> **STOP and Think:** After looking at the tree in Figure 7.42, where do you think all humans came from?

If humans populated Africa before Asia, then African genomes started to diverge from each other earlier than Asian genomes. Thus, we would expect to find that African genomes, which had more time to diverge from each other, have more mutations (compared to each other and other genomes) than Asian genomes. This reasoning gives us a hint on how to check whether the human race spread from Africa.

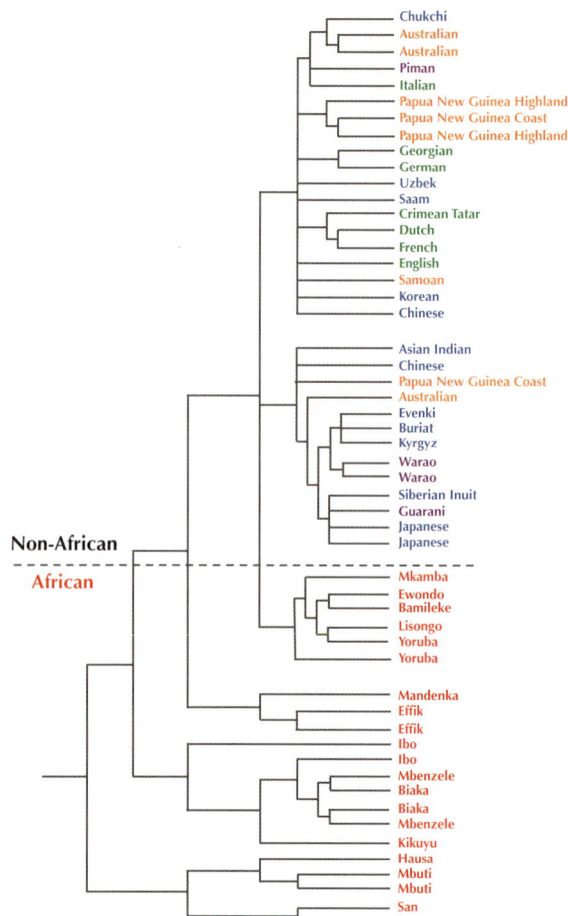

FIGURE 7.42 Evolutionary tree constructed for various human mitochondrial genomes from Africa (red), Asia (blue), North and South America (purple), Europe (green), and Oceania (orange). The dashed line divides African from non-African genomes.

African genomes are indeed more diverse than genomes from other continents, which led Wilson and colleagues to conclude that the African lineage is the oldest and that modern humans trace their roots back to Africa. Thus, a population of Africans, the first modern humans, forms one subtree, whereas another subtree represents a subgroup that left Africa and later spread out to the rest of the world. Using the mitochondrial tree, Wilson and colleagues further estimated that humans emerged from Africa 130,000 years ago, with racial differences arising only 50,000 years ago. Figure 7.43 shows putative human migration patterns derived from genomic data.

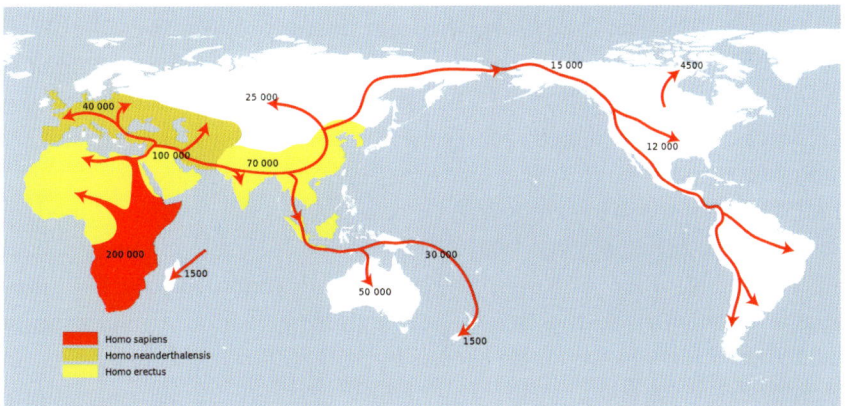

FIGURE 7.43 Putative routes of human migration derived from genomic data labeled by the number of years ago these migrations occurred.

CHAPTER 7

Bibliography Notes

Zuckerkandl and Pauling, 1965 published "Molecules as documents of evolutionary history". The evolutionary tree that resolved the giant panda riddle was constructed by O'Brien et al., 1985. The *Out of Africa* hypothesis was proposed by Cann, Stoneking, and Wilson, 1987. The studies of the origins of HIV and the primate reservoir of the HIV virus were initiated by Gao et al., 1999. The first molecular evidence of HIV transmission in a criminal case was presented by Metzker et al., 2002. Whiting, Bradler, and Maxwell, 2003 published the study of loss and recovery of wings in stick insects.

The UPGMA approach to evolutionary tree reconstruction was developed by Sokal and Michener, 1958. The neighbor joining algorithm was developed by Saitou and Nei, 1987. Studier and Keppler, 1988 proved that this algorithm solves the Distance-Based Phylogeny Problem for additive trees. The dynamic programming algorithm solving the Small Parsimony problem was developed by Sankoff, 1975. The four point condition was formulated by Zaretskii, 1965. The nearest neighbor interchange approach to exploring trees was proposed by Robinson, 1971. Felsenstein, 2004 provides excellent coverage of various algorithms for tree reconstruction.

WHICH ANIMAL GAVE US SARS?

HOW DID YEAST BECOME A WINEMAKER?
Clustering Algorithms

An Evolutionary History of Wine Making

How long have we been addicted to alcohol?

One of the first organisms that humans domesticated was yeast. In 2011, while excavating an old graveyard in an Armenian cave, scientists discovered a 6,000 year-old winery, complete with a wine press, fermentation vessels, and even drinking cups. This winery was a major technological innovation that required understanding how to control *Saccharomyces*, the genus of yeast used in alcohol and bread production.

Yet our interest in alcohol may date back much farther than 6,000 years. In 2008, scientists discovered that pen-tailed tree shrews (Figure 8.1), which are similar to the ancient ancestors of all primates, are alcoholics. Their beverage of choice? A "palm wine" produced by the flowers of Bertram palms and naturally fermented by *Saccharomyces* yeast living on the flowers. This finding suggests that our own taste for alcohol may have a genetic basis that predates the Armenian wine cave by millions of years!

FIGURE 8.1 The pen-tailed tree shrew.

Pound for pound, the amount of palm wine that tree shrews consume would be lethal to most mammals. Fortunately, the shrews have developed efficient ways of metabolizing alcohol, and so they avoid inebriation, which would increase their risk of being killed by a predator. Because of tree shrews' alcohol tolerance, scientists believe that alcohol probably offers the shrews some evolutionary advantages, such as protection against heart attack. It is also possible that our more recent primate ancestors were also heavy drinkers — after all, chimpanzees drink naturally brewed fruit nectar — and that we may have inherited an ancestral association of alcohol intake with caloric gain.

Chapter 8

The diauxic shift

The species of yeast that we will consider in this chapter is *Saccharomyces cerevisiae*, which can brew wine because it converts the **glucose** found in fruit into **ethanol**. We will therefore begin with a simple question: if *S. cerevisiae* often lives on grapevines, why must crushed grapes be stored in tightly sealed barrels in order to make wine?

Once its supply of glucose runs out, *S. cerevisiae* must do something to survive. It therefore inverts its metabolism, with the ethanol that it just produced becoming its new food supply. This metabolic inversion, called the **diauxic shift**, can only occur in the presence of oxygen. Without oxygen, *S. cerevisiae* hibernates until either glucose or oxygen becomes available. In other words, if winemakers don't seal their barrels, then the yeast in the barrel will metabolize the ethanol that it just produced, ruining the wine.

The diauxic shift is a complex process that affects the expression of many genes. Accordingly, it must derive from a major evolutionary event that equipped the *Saccharomyces* ancestor with a formidable advantage over its competitors: not only can *Saccharomyces* kill its competitors by producing ethanol, which is toxic to most bacteria and other yeasts, but it can then use the accumulated ethanol as an energy source. But how and when did *Saccharomyces* invent the diauxic shift? And which genes does it involve?

Identifying Genes Responsible for the Diauxic Shift

Two evolutionary hypotheses with different fates

Remember Susumu Ohno and his Random Breakage Model? Ohno also hypothesized that there exist rare evolutionary events called **whole genome duplications**, or **WGDs**, which duplicate an entire genome. He proposed this **WGD Model** in 1970, when there was absolutely no evidence to support it. Yet he believed that a WGD would be needed at the time of a critical evolutionary innovation for a species to implement some revolutionary new function, such as the diauxic shift.

For example, imagine the time, millions of years ago, when the first fruit-bearing plants had evolved, but no organisms could metabolize the glucose produced by these fruits and take advantage of the produced ethanol. The first species to do so would have possessed an enormous evolutionary advantage, but metabolizing glucose — let alone ethanol — is not a simple task. Rather than creating a new gene here or there, the diauxic shift would have required creating new metabolic pathways with many genes working together. Ohno argued that a WGD would provide a platform for such a revolutionary innovation, since every duplicated gene would have two copies. One copy would be free to evolve without compromising the gene's existing function, which would be carried out by the remaining copy.

The Random Breakage Model and the WGD Model had very different fates. From the time of its proposal, the Random Breakage Model was embraced by biologists, and it became dogma until its refutation in 2003. In contrast, the WGD Model was initially met with skepticism (because only 13% of *S. cerevisiae* genes are duplicated) and did not gain traction for 25 years.

In 1997, Wolfe and Shields provided the first computational arguments in favor of a WGD in *S. cerevisiae*. They argued that the fact that only 13% of *S. cerevisiae* genes are duplicated is not surprising because even if hundreds of genes contribute to an evolutionary innovation after a WGD, most genes are not needed for this innovation. Thus, unneeded duplicate genes will be bombarded by mutations until they turn into pseudogenes and eventually disappear from the genome after millions of years. See **DETOUR: Whole Genome Duplication or a Series of Single-Gene Duplications?** to **PAGE 111** learn more about the arguments surrounding a WGD in *S. cerevisiae*.

Which yeast genes drive the diauxic shift?

One of the many steps that yeast performs during fermentation is the conversion of **acetaldehyde** into ethanol. If oxygen becomes subsequently available, then the accumulated ethanol is converted back into acetaldehyde. Both the acetaldehyde-to-ethanol and ethanol-to-acetaldehyde conversions are catalyzed by an enzyme called **alcohol dehydrogenase (Adh)**. In *S. cerevisiae*, Adh activity is encoded by two genes, Adh_1 and Adh_2, which arose from the duplication of a single ancestral gene. The enzyme encoded by Adh_1 has an elevated ability of producing ethanol, whereas the enzyme encoded by Adh_2 has an elevated ability of consuming ethanol.

In 2005, Michael Thomson used a multiple alignment of *Adh* genes from various yeast species to reconstruct an ancient *Saccharomyces* gene. This gene showed a preference to convert acetaldehyde to ethanol, which resembled the behavior of Adh_1. Thomson therefore concluded that before the WGD in *Saccharomyces*, alcohol dehydrogenase was mainly involved in the generation, not consumption, of ethanol. After the WGD, Adh_1 carried out its original function, whereas Adh_2 was free to help power the diauxic shift.

STOP and Think: How would you find the rest of the genes in *S. cerevisae* that work together to accomplish the diauxic shift?

CHAPTER 8

Imagine that you were able to monitor all n yeast genes at m time checkpoints on either side of the diauxic shift, resulting in an $n \times m$ **gene expression matrix** E, where $E_{i,j}$ is a number representing the expression level of gene i at checkpoint j. The i-th row of E is called the **expression vector** of gene i. Just by looking at the yeast genes' expression vectors, you would observe different patterns of gene behavior with respect to the diauxic shift. You would see genes whose expression hardly changes, genes whose expression rapidly increases before the diauxic shift and decreases afterwards, genes whose expression suddenly increases after the diauxic shift, and so on.

Although this chapter focuses on gene expression with respect to the diauxic shift, expression matrices are commonplace in biological analysis. For example, if the expression vector of a newly sequenced gene is similar to the expression vector of a gene with known function, a biologist may suspect that these genes perform related functions. Also, genes with similar expression vectors may imply that the genes are co-regulated, meaning that their expression is controlled by the same transcription factor. This suggests a "guilt by association" strategy for inferring gene functions by starting from a few genes with known functions and potentially propagating the functions of these genes to other genes with similar expression vectors.

Finally, gene expression analysis is important in biomedical studies such as analyzing tissues before and after a drug is administered or contrasting cancerous and non-cancerous cells. For example, expression analysis led to **MammaPrint**, a diagnostic test that determines the likelihood of breast cancer recurrence based on the expression analysis of 70 human genes associated with tumor activation and suppression.

Yet across all these applications, the question remains: what methods do biologists use to analyze gene expression data?

Introduction to Clustering

Gene expression analysis

In 1997, Joseph DeRisi conducted the first massive gene expression experiment by sampling an *S. cerevisiae* culture seven times at hours -6, -4, -2, 0, +2, +4, and +6, where hour 0 indicates the diauxic shift. Since there are approximately 6,400 genes in *S. cerevisiae*, this experiment resulted in a $6{,}400 \times 7$ gene expression matrix.

 **STOP and Think:** What technology would you use to generate this matrix?

HOW DID YEAST BECOME A WINE MAKER?

We have already encountered three technologies that could be used to generate this matrix (see DETOUR: Measuring Gene Expression), but none of these technologies had matured by 1997! For this reason, DeRisi had to use **microarrays**, which differ from the DNA arrays that we discussed in Chapter 2 (see DETOUR: Microarrays). Microarrays are rarely used today, but the algorithmic approaches deRisi used for microarray analysis work equally well for modern gene expression technologies.

PAGE 111

PAGE 112

> **STOP and Think:** Figure 8.2 visualizes the expression vectors of three yeast genes. Which of these genes do you think are involved in the diauxic shift?

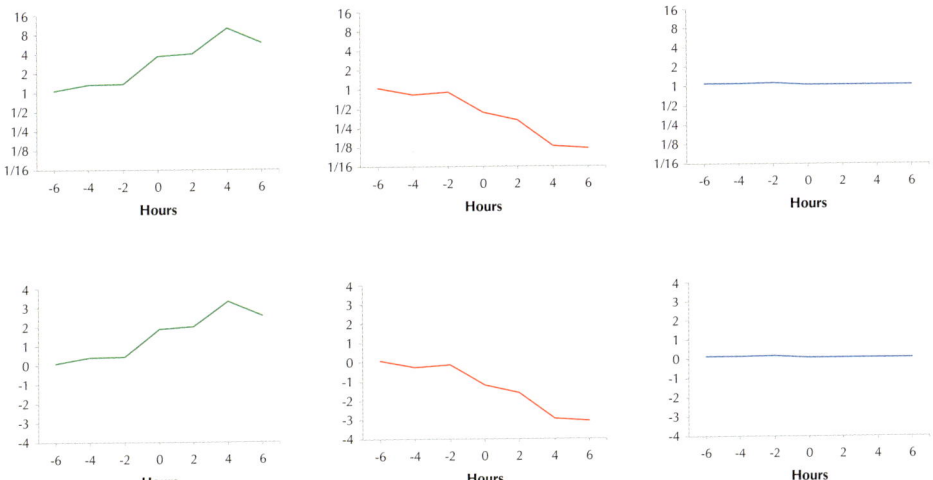

FIGURE 8.2 (Top) Expression vectors $(1.07, 1.35, 1.37, 3.70, 4.00, 10.00, 5.88)$, $(1.06, 0.83, 0.90, 0.44, 0.33, 0.13, 0.12)$, and $(1.11, 1.11, 1.12, 1.06, 1.05, 1.06, 1.05)$ of three yeast genes (YLR258W, YPL012W, and YPR055W, respectively) visualized as plots. Each expression vector $(e_1, \ldots, e_m)$ is represented as a collection of line segments connecting points (j, e_j) to $(j+1, e_{j+1})$ for each j between 1 and $m-1 = 6$. In the DeRisi experiment, the expression level at the initial checkpoint corresponds to the base level of expression; note that it is close to 1 for the three genes. Values above 1 in expression vectors correspond to increased expression, while values below 1 correspond to decreased expression. (Bottom) The expression vectors of the same three genes with expression levels substituted by their base-2 logarithms: $(0.11, 0.43, 0.45, 1.89, 2.00, 3.32, 2.56)$, $(0.09, -0.28, -0.15, -1.18, -1.59, -2.96, -3.08)$, and $(0.15, 0.15, 0.17, 0.09, 0.07, 0.09, 0.07)$.

CHAPTER 8

Note that the pattern of the expression vector of gene YPR055W (Figure 8.2 (top left)) remains flat during the diauxic shift. We therefore conclude that this gene is probably not involved in the diauxic shift. On the other hand, the expression of gene YLR258W (Figure 8.2 (top right)) significantly changes during the diauxic shift, leading us to hypothesize that this gene is involved in the diauxic shift. Indeed, checking the *Saccharomyces* Genome Database (http://yeastgenome.org) reveals that YLR258W is **glycogen synthase**. This enzyme controls the production of **glycogen**, a glucose polysaccharide that is the main storage vessel for glucose in yeast cells.

In practice, biologists often take the logarithm of expression values (Figure 8.2 (bottom)). After this transformation, positive values of a gene's expression vector correspond to increased expression, and negative values correspond to decreased expression. Figure 8.3 shows an expression matrix of ten yeast genes after taking logarithms.

Gene	Expression Vector						
YLR361C	0.14	0.03	-0.06	0.07	-0.01	-0.06	-0.01
YMR290C	0.12	-0.23	-0.24	-1.16	-1.40	-2.67	-3.00
YNR065C	-0.10	-0.14	-0.03	-0.06	-0.07	-0.14	-0.04
YGR043C	-0.43	-0.73	-0.06	-0.11	-0.16	3.47	2.64
YLR258W	0.11	0.43	0.45	1.89	2.00	3.32	2.56
YPL012W	0.09	-0.28	-0.15	-1.18	-1.59	-2.96	-3.08
YNL141W	-0.16	-0.04	-0.07	-1.26	-1.20	-2.82	-3.13
YJL028W	-0.28	-0.23	-0.19	-0.19	-0.32	-0.18	-0.18
YKL026C	-0.19	-0.15	0.03	0.27	0.54	3.64	2.74
YPR055W	0.15	0.15	0.17	0.09	0.07	0.09	0.07

FIGURE 8.3 A 10×7 submatrix of DeRisi's $6{,}400 \times 7$ gene expression matrix for ten yeast genes (after taking the base-2 logarithm of each expression value). The genes from Figure 8.2 are colored appropriately.

Clustering yeast genes

Our goal is to **partition** the set of all yeast genes into k disjoint **clusters** so that genes in the same cluster have similar expression vectors. In practice, the number of clusters is not known *a priori*, and so biologists typically apply clustering algorithms to gene expression data for various values of k, selecting the value of k that makes sense biologically. For simplicity, we will assume that k is fixed. Figure 8.4 shows a partition of the genes from Figure 8.3 into three clusters indicating increased, decreased, and flat expression during the diauxic shift.

Although the diauxic shift is an important event in the life of *S. cerevisiae*, it has no bearing on most of the yeast's functions. We therefore suspect that most *S. cerevisiae* genes have flat expression during the diauxic shift, and we would like to exclude these genes from further consideration, thus reducing the size of the expression matrix.

Gene	Expression Vector						
YLR361C	**0.14**	0.03	-0.06	0.07	-0.01	-0.06	-0.01
YMR290C	0.12	-0.23	-0.24	-1.16	-1.40	-2.67	**-3.00**
YNR065C	-0.10	**-0.14**	-0.03	-0.06	-0.07	-0.14	-0.04
YGR043C	-0.43	-0.73	-0.06	-0.11	-0.16	**3.47**	2.64
YLR258W	0.11	0.43	0.45	1.89	2.00	**3.32**	2.56
YPL012W	0.09	-0.28	-0.15	-1.18	-1.59	-2.96	**-3.08**
YNL141W	-0.16	-0.04	-0.07	-1.26	-1.20	-2.82	**-3.13**
YJL028W	-0.28	-0.23	-0.19	-0.19	**-0.32**	-0.18	-0.18
YKL026C	-0.19	-0.15	0.03	0.27	0.54	**3.64**	2.74
YPR055W	0.15	0.15	**0.17**	0.09	0.07	0.09	0.07

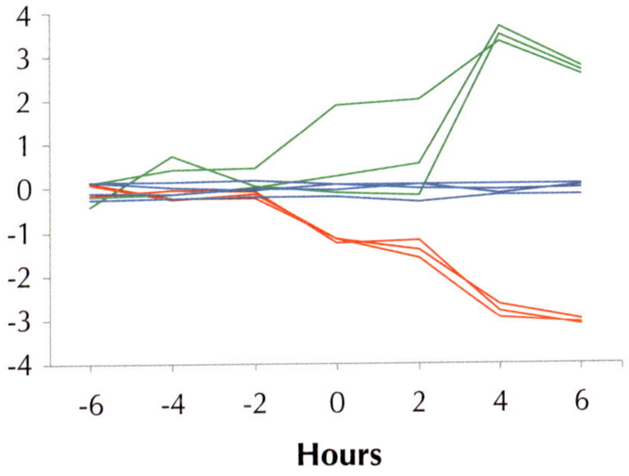

FIGURE 8.4 The rows of the gene expression matrix from Figure 8.3 partitioned into three clusters. Green genes exhibit increased expression, red genes exhibit decreased expression, and blue genes exhibit flat behavior and are unlikely to be associated with the diauxic shift. The element with the largest absolute value in each expression vector is shown in bold. (Bottom) The rows of the matrix visualized as plots.

The expression levels of most yeast genes hardly change before and after the diauxic shift (the blue genes in Figure 8.4). These genes also have the property that all of their expression vector values are very close to zero. In our analysis, we will exclude genes with expression vectors whose values are all between -2.3 and 2.3. Doing so reduces the

CHAPTER 8

original 6,400 yeast gene dataset to a dataset containing 230 genes whose expression changes significantly around the diauxic shift.

You can see in Figure 8.4 that gene YLR258W has a different pattern of change than gene YGR043C, indicating that dividing the 230 yeast genes into just two clusters (i.e., those with increasing and decreasing expression levels) may be too simplistic. Our goal is to cluster these genes based on similar patterns of behavior.

The Good Clustering Principle

To identify groups of genes with similar expression patterns, we will think of an expression vector of length m as a point in m-dimensional space; genes with similar expression vectors will therefore form clusters of nearby points. Ideally, clusters should satisfy the following common-sense principle, which is illustrated in Figure 8.5.

Good Clustering Principle: *Every pair of points from the same cluster should be closer to each other than any pair of points from different clusters.*

We have therefore embedded gene expression analysis within the algorithmic problem of partitioning a collection of n points in m-dimensional space into k clusters, which will be our focus in this chapter.

Good Clustering Problem:
Partition a set of points into clusters.

> **Input**: A set of n points in m-dimensional space and an integer k.
> **Output**: A partition of the n points into k clusters satisfying the Good Clustering Principle.

EXERCISE BREAK: Form ten points in two-dimensional space by taking the fourth and seventh columns of the matrix in Figure 8.4. How should these points be partitioned into three clusters?

EXERCISE BREAK: Compute the number of partitions of n points into two non-empty clusters.

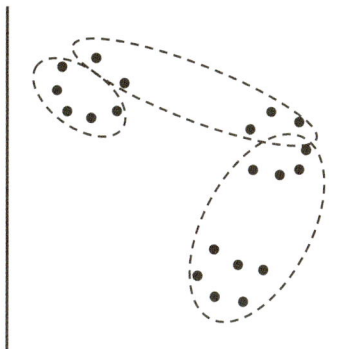 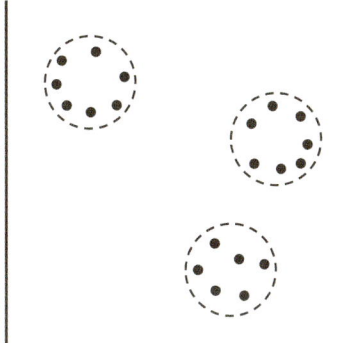

FIGURE 8.5 (Left) A partition of twenty points into three clusters that do not satisfy the Good Clustering Principle. (Right) A different partition of these points that does satisfy the Good Clustering Principle.

The eye naturally divides the points in Figure 8.6 (left) into two clusters. Unfortunately, these clusters do not satisfy the Good Clustering Principle; in fact, no such partition of these points into two clusters exists! As a result, we will need to take a different approach in order to devise a well-defined computational problem for clustering.

EXERCISE BREAK: Design a polynomial algorithm to check whether there is a solution of the Good Clustering Problem.

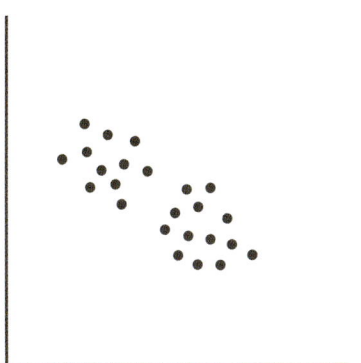 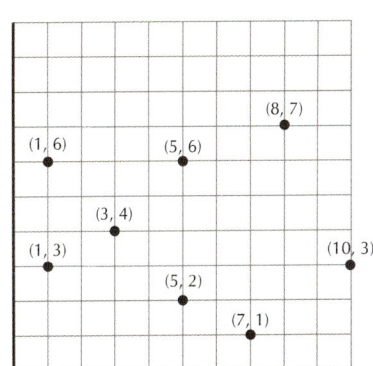

FIGURE 8.6 (Left) A collection of points that obviously form two clusters but for which no partition into two clusters satisfies the Good Clustering Principle. (Right) Eight points in two-dimensional space.

CHAPTER 8

> **STOP and Think:** Figure 8.6 (right) shows eight data points in two-dimensional space. How would you partition these points into three clusters? How can we transform the Good Clustering Problem into a well-defined computational problem?

Clustering as an Optimization Problem

Rather than thinking about clustering as dividing data points *Data* into *k* clusters, we will instead try to select a set *Centers* of *k* points that will serve as the **centers** of these clusters. We would like to choose *Centers* so that they minimize some distance function between *Centers* and *Data* over all possible choices of centers. But how should this distance function be defined?

First, we define the **Euclidean distance** between points $v = (v_1, \ldots, v_m)$ and $w = (w_1, \ldots, w_m)$ in *m*-dimensional space, denoted $d(v, w)$, as the length of the line segment connecting these points,

$$d(v, w) = \sqrt{\sum_{i=1}^{m}(v_i - w_i)^2}.$$

Next, given a point *DataPoint* in multi-dimensional space and a set of *k* points *Centers*, we define the distance from *DataPoint* to *Centers*, denoted $d(DataPoint, Centers)$, as the Euclidean distance from *DataPoint* to its closest center,

$$d(DataPoint, Centers) = \min_{\text{all points } x \text{ from } Centers} d(DataPoint, x).$$

The length of the segments in Figure 8.7 correspond to $d(DataPoint, Centers)$ for each point *DataPoint*.

We now define the distance between all data points *Data* and centers *Centers*. This distance, denoted MAXDISTANCE(*Data*, *Centers*), is the maximum of $d(DataPoint, Centers)$ among all data points *DataPoint*,

$$\text{MAXDISTANCE}(Data, Centers) = \max_{\text{all points } DataPoint \text{ from } Data} d(DataPoint, Centers).$$

In Figure 8.7, this distance corresponds to the length of the red segment. We can now formulate a well-defined clustering problem.

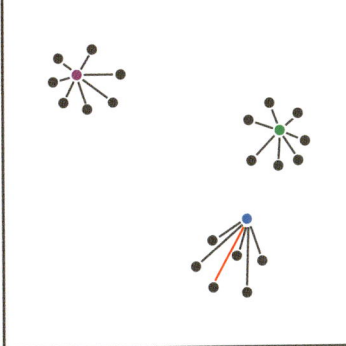

FIGURE 8.7 The collection of points *Data* from Figure 8.5 (shown as black points) along with three centers forming a set *Centers* (shown as colored points). For each point *DataPoint* in *Data*, d(*DataPoint, Centers*) is equal to the length of the segment connecting it to its nearest center. MAXDISTANCE(*Data, Centers*) is equal to the length of the longest such segment, which is shown in red.

k-Center Clustering Problem:

Given a set of data points, find k centers minimizing the maximum distance between these data points and centers.

Input: A set of points *Data* and an integer k.
Output: A set *Centers* of k centers that minimize the distance MAXDISTANCE(*DataPoints, Centers*) over all possible choices of k centers.

EXERCISE BREAK: How would you select a center in the case of only a single cluster (i.e., when $k = 1$)?

Farthest First Traversal

Although the k-Center Clustering Problem is easy to state, it is *NP*-Hard. The **Farthest First Traversal** heuristic, whose pseudocode is shown below, selects centers from the points in *Data* (instead of from all m-dimensional points). It first selects an arbitrary point in *Data* as the first center and iteratively adds a new center as the point in *Data* that is farthest from the centers chosen so far, with ties broken arbitrarily (Figure 8.8).

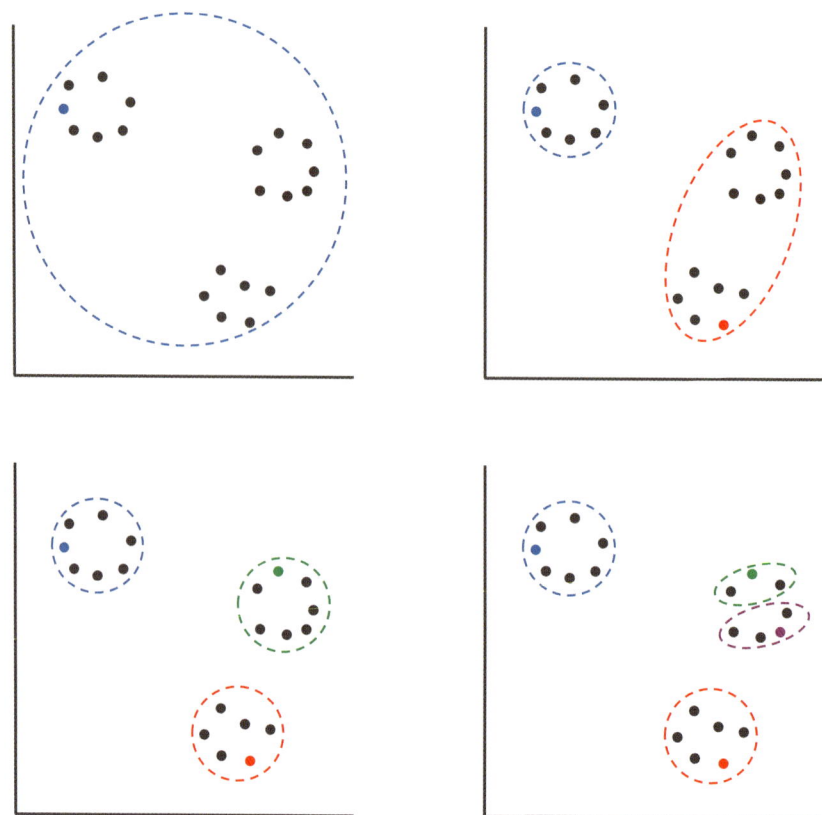

FIGURE 8.8 Applying FARTHESTFIRSTTRAVERSAL to the data in Figure 8.5. (Top left) An arbitrary point from the dataset (shown in blue) is selected as the first center. All points belong to a single cluster. (Top right) The red point is selected as the second center, since it is the farthest from the blue point. (Bottom left) After computing each data point's minimum distance to each of the first two centers, we find that the point with the largest such distance is the green point, which becomes the third center. (Bottom right) The fourth center is shown in purple.

FARTHESTFIRSTTRAVERSAL(*Data*, *k*)
    Centers ← the set consisting of a single randomly chosen point from *Data*
    while |*Centers*| < *k*
        DataPoint ← the point in *Data* maximizing *d*(*DataPoint*, *Centers*)
        add *DataPoint* to *Centers*
    return *Centers*

EXERCISE BREAK: Apply FARTHESTFIRSTTRAVERSAL to the eight points in Figure 8.6 (right) with $k = 3$. How does the result change if you change the point that is selected first?

EXERCISE BREAK: Let *Centers* be the set of centers returned by FARTHESTFIRSTTRAVERSAL, and let *Centers*$_{opt}$ be a set of centers corresponding to an optimal solution of the *k*-Center Clustering Problem. Prove that

$$\text{MAXDISTANCE}(Data, Centers) \leq 2 \cdot \text{MAXDISTANCE}(Data, Centers_{opt}).$$

Can you find a collection of data points such that the centers returned by FARTHESTFIRSTTRAVERSAL are suboptimal?

FARTHESTFIRSTTRAVERSAL is fast, and according to the preceding exercise, its solution approximates the optimal solution of the *k*-Center Clustering Problem; however, this algorithm is rarely used for gene expression analysis. In *k*-Center Clustering, we selected *Centers* so that these points would minimize MAXDISTANCE(*Data, Centers*), the maximum distance between any point in *Data* and its nearest center. But biologists are usually interested in analyzing *typical* rather than *maximum* deviations, since the latter may correspond to outliers representing experimental errors (Figure 8.9).

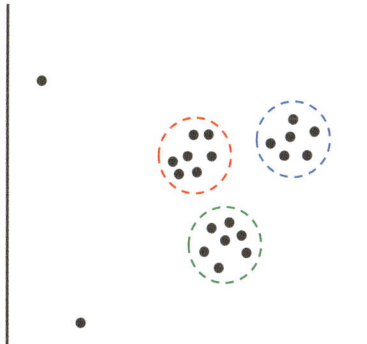
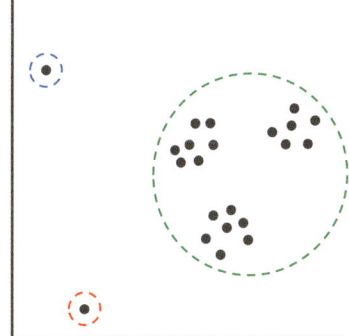

FIGURE 8.9 (Left) A set of data points with three clearly seen clusters and two outliers. (Right) Because FARTHESTFIRSTTRAVERSAL relies on MAXDISTANCE to compute new centers, if we attempt to cluster the data points into three clusters, then regardless of which point is selected as the first center, the two outliers on the left will be selected as centers of single-element clusters, and all remaining data points will be assigned to a single cluster.

CHAPTER 8

 **STOP and Think:** Can you devise an alternative scoring function that is more biologically appropriate than MAXDISTANCE(*Data, Centers*)?

k-Means Clustering

Squared error distortion

To address limitations of MAXDISTANCE, we will introduce a new scoring function. Given a set *Data* of *n* data points and a set *Centers* of *k* centers, the **squared error distortion** of *Data* and *Centers*, denoted DISTORTION(*Data, Centers*), is defined as the mean squared distance from each data point to its nearest center,

$$\text{DISTORTION}(Data, Centers) = \frac{1}{n} \sum_{\text{all points } DataPoint \text{ in } Data} d(DataPoint, Centers)^2.$$

Note that whereas MAXDISTANCE(*Data, Centers*) only accounts for the length of the single red segment in Figure 8.7, the squared error distortion accounts for the length of all segments in this figure.

 EXERCISE BREAK: Compute the values of MAXDISTANCE(*Data, Centers*) and DISTORTION(*Data, Centers*) for the eight data points and the three centers shown in Figure 8.10 (left). How do these values differ for the centers in Figure 8.10 (right)?

 **Squared Error Distortion Problem**:

Compute the squared error distortion of a set of data points with respect to a set of centers.

Input: A set of points *Data* and a set of centers *Centers*.
Output: The squared error distortion DISTORTION(*Data, Centers*).

The squared error distortion leads us to the following modification of the *k*-Centers Clustering Problem.

HOW DID YEAST BECOME A WINE MAKER?

k-Means Clustering Problem:

Given a set of data points, find k center points minimizing the squared error distortion.

Input: A set of points *Data* and an integer k.
Output: A set *Centers* of k centers that minimize DISTORTION(*Data*, *Centers*) over all possible choices of k centers.

Although the k-Centers and k-Means Clustering Problems look similar, they may produce different results (Figure 8.10). The key difference between the k-Centers and k-Means Clustering Problems is that in the latter, the placement of a center is far less affected by outliers (Figure 8.11).

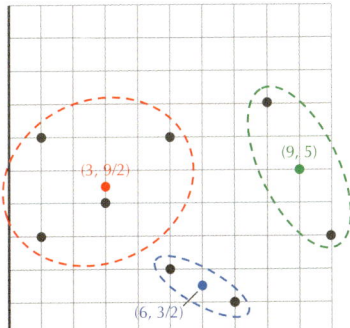 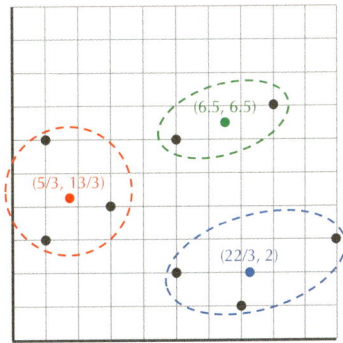

FIGURE 8.10 (Left) The three colored centers solving the k-Centers Clustering Problem for the eight black data points from Figure 8.6 (right), along with the clusters that they form. (Right) The three colored centers solving the k-Means Clustering Problem for these points, along with the clusters that they form.

k-means clustering and the center of gravity

It turns out that the k-Means Clustering Problem is *NP*-Hard when $k > 1$. However, when $k = 1$, the k-Means Clustering Problem amounts to finding a single center point x that minimizes the squared error distortion. Although we acknowledge that partitioning a set of data points into a single cluster is trivial, it remains unclear how to find a single center minimizing the squared error distortion. We would like to solve this simpler problem because it will help us design a heuristic for the case when $k > 1$.

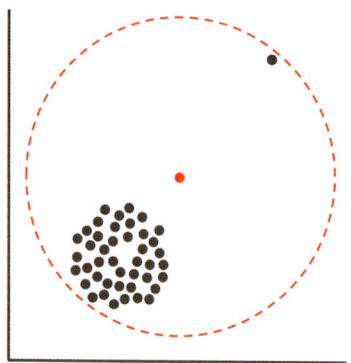
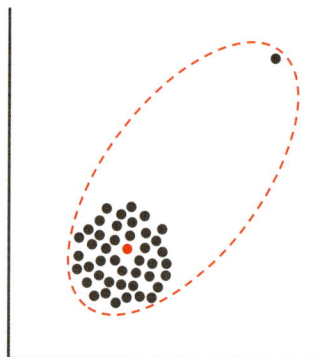

Figure 8.11 Center placement varies in different clustering problem formulations. (Left) In the *k*-Center Clustering Problem, a cluster's center is chosen so that the maximum distance between the center and any point in the cluster is minimized. As a result, the position of the center can be greatly influenced by outliers. (Right) In the *k*-Means Clustering Problem, the outlier's influence over the placement of the center is much smaller. This behavior is preferable when analyzing biological datasets, in which outliers often correspond to erroneous data.

We define the **center of gravity** of *Data* as the point whose *i*-th coordinate is the average of the *i*-th coordinates of all points from *Data*. For example, the center of gravity of the points $(3, 8)$, $(8, 0)$, and $(7, 4)$ is

$$\left(\frac{3+8+7}{3}, \frac{8+3+4}{3} \right) = (6, 5).$$

Center of Gravity Theorem: *The center of gravity of a set of points Data is the unique point solving the k-Means Clustering Problem for* $k = 1$.

For a proof of this theorem, see **DETOUR: Proof of the Center of Gravity Theorem**.

EXERCISE BREAK: Although the *k*-Means Clustering Problem is *NP*-hard for $k > 1$, it can be solved in polynomial time for any value of *k* in the case of clustering in one-dimensional space, i.e., when all data points fall on a line. Design an algorithm for solving the *k*-Means Clustering Problem in this case.

EXERCISE BREAK: Prove that the centers in Figure 8.10 (right) solve the *k*-Means Clustering Problem for $k = 3$.

The Lloyd Algorithm

From centers to clusters and back again

The **Lloyd algorithm** is one of the most popular clustering heuristics for the *k*-Means Clustering Problem. It first chooses *k* arbitrary points *Centers* from *Data* as centers and then iteratively performs the following two steps (Figure 8.12):

- **Centers to Clusters:** After centers have been selected, assign each data point to the cluster corresponding to its nearest center; ties are broken arbitrarily.
- **Clusters to Centers:** After data points have been assigned to clusters, assign each cluster's center of gravity to be the cluster's new center.

STOP and Think: Is it possible for the Lloyd algorithm to produce two centers that coincide (thus resulting in fewer than *k* clusters)?

In Figure 8.12, the centers appear to be moving less and less between iterations. We say that the Lloyd algorithm has **converged** if the centers (and therefore their clusters) stop changing between iterations.

STOP and Think: Can you find a set of data points for which the Lloyd algorithm does not converge?

If the Lloyd algorithm has not converged, the squared error distortion must decrease in any step, according to the following reasoning:

- In a "Centers to Clusters" step, if a data point is assigned to a new center, then this point must be closer to the new center than its previous center. Thus, the squared error distortion must decrease.
- In a "Clusters to Center" step, if a center is updated as a cluster's center of gravity, then by the Center of Gravity Theorem, the new center is the *only* point minimizing the squared error distortion for the points in its cluster. Thus, the squared error distortion must decrease.

STOP and Think: Does this reasoning imply that the Lloyd algorithm must converge?

CHAPTER 8

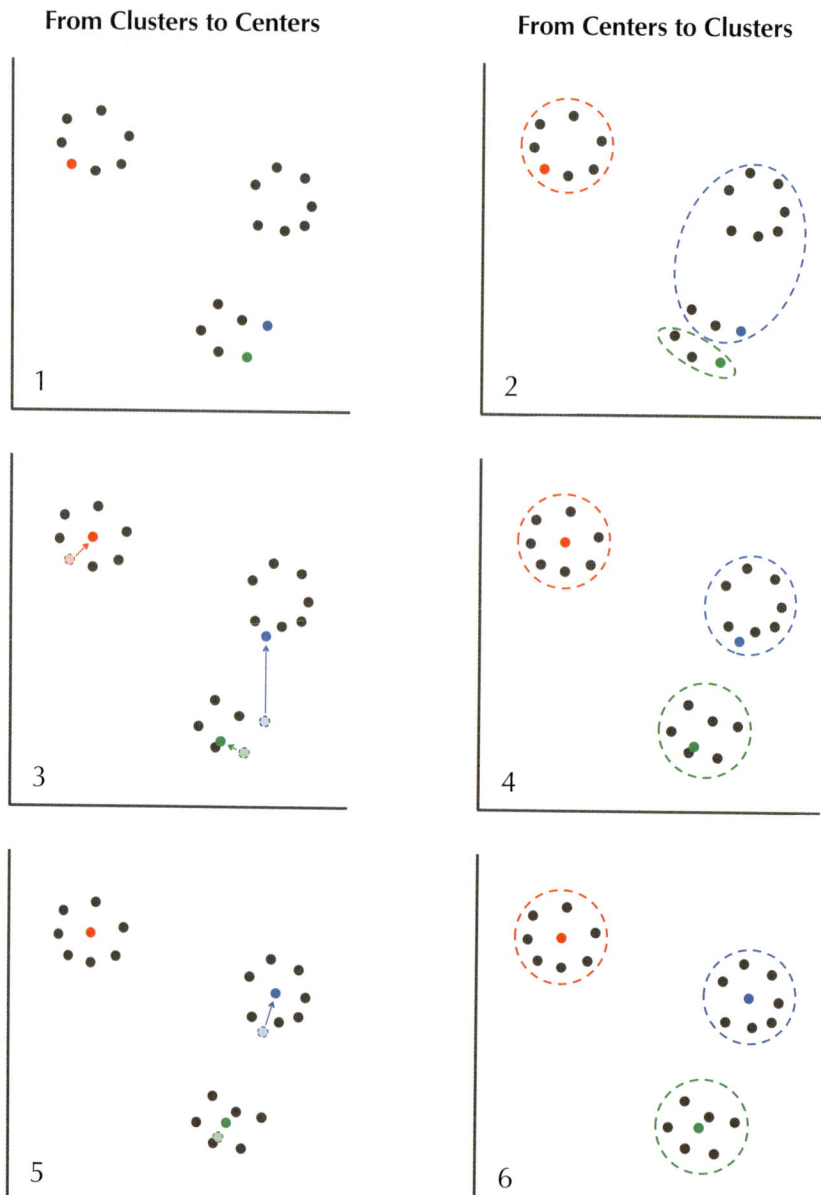

FIGURE 8.12 The Lloyd algorithm in action for $k = 3$. In the top left panel, we select three arbitrary data points as centers, shown as differently colored points. In subsequent panels, we iterate the "Centers to Clusters" step followed by the "Clusters to Centers" step. In the bottom right panel, the Lloyd algorithm has converged.

It is not true that the Lloyd algorithm must converge just because the squared error distortion decreases at each step. For example, it could be the case that subsequent decreases in squared error distortion become smaller and smaller, leading to an infinite process (e.g., if the error distortion decreases by 1/2, then 1/4, then 1/8, and so on). The following exercise ensures that such a scenario cannot occur.

EXERCISE BREAK: Prove that the number of iterations of the Lloyd algorithm does not exceed the number of partitions of the data points into k clusters.

EXERCISE BREAK: Pick your favorite parameter k and run the Lloyd algorithm 1,000 times on the 230-gene diauxic shift dataset, each time initialized with a new set of k randomly chosen centers. Construct a histogram of the squared error distortions of the resulting 1,000 outcomes. How many times did you have to run the Lloyd algorithm before finding the run that scored highest among your 1,000 runs?

Initializing the Lloyd algorithm

Figure 8.13 illustrates that things can go horribly wrong if we do not pay attention to the Lloyd algorithm's initialization step. In Figure 8.13 (top), we select no centers from clump 1, two centers from clump 3, and one center from each of clumps 2, 4, and 5. As shown in Figure 8.13 (bottom), after the first iteration of the Lloyd algorithm, all points in clumps 1 and 2 will be assigned to the red center, which will move approximately halfway between clumps 1 and 2. The two centers in clump 3 will divide the points in that clump into two clusters. And the centers in clumps 4 and 5 will move toward the middle of these clumps. The Lloyd algorithm will then quickly converge, resulting in an incorrect clustering.

EXERCISE BREAK: Compute the probability that at least one of the five clumps in Figure 8.13 will have no centers if five centers are chosen randomly from the data (like in the Lloyd algorithm).

STOP and Think: How would you change the Lloyd algorithm's initialization step to improve the clusters it finds?

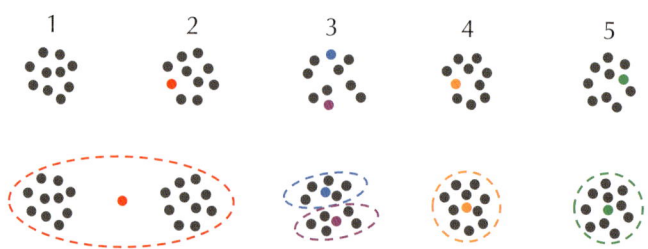

Figure 8.13 (Top) Five clumps of ten points in two-dimensional space. The Lloyd algorithm is initialized so that clump 1 contains no centers, clump 3 contains two centers (blue and purple), and each of the other three clumps contains one center (red, orange, and green). (Bottom) The Lloyd algorithm has combined the points in clumps 1 and 2 into a single cluster and split clump 3 into two clusters.

k-means++ Initializer

We have thus far not paid much attention to how initial centers are chosen in the Lloyd algorithm, which selects them randomly. Similarly to FARTHESTFIRSTTRAVERSAL, *k*-MEANS++INITIALIZER picks *k* centers one at a time, but instead of choosing the point farthest from those picked so far, it chooses each point at random in such a way that distant points are more likely to be chosen than nearby points. Specifically, the probability of selecting a center *DataPoint* from *Data* is proportional to the squared distance of *DataPoint* from the centers already chosen, i.e., to $d(DataPoint, Centers)^2$.

For a simple example, say that we have just three data points, and that the squared distances from these points to the existing centers *Centers* are equal to 1, 4, and 5. Then the probability of *k*-MEANS++INITIALIZER selecting each of these points as the next center is 1/10, 4/10, and 5/10, respectively.

---

k-MEANS++INITIALIZER(*Data*, *k*)
    Centers ← the set consisting of a single randomly chosen point from *Data*
    while |*Centers*| < *k*
        randomly select *DataPoint* from *Data* with probability proportional to
            $d(DataPoint, Centers)^2$
        add *DataPoint* to *Centers*
    return *Centers*

STOP and Think: Although *k*-MEANS++INITIALIZER may also fall into the trap in Figure 8.13, it does so rarely. Why?

HOW DID YEAST BECOME A WINE MAKER?

EXERCISE BREAK: Power up your implementation of the Lloyd algorithm using *k*-MEANS++INITIALIZER, and apply it to the 230-gene diauxic shift dataset for varying values of k.

Clustering Genes Implicated in the Diauxic Shift

Since selecting the most biologically relevant value of k can be challenging, we will (somewhat arbitrarily) choose to cluster the 230 yeast genes into six clusters (Figure 8.14).

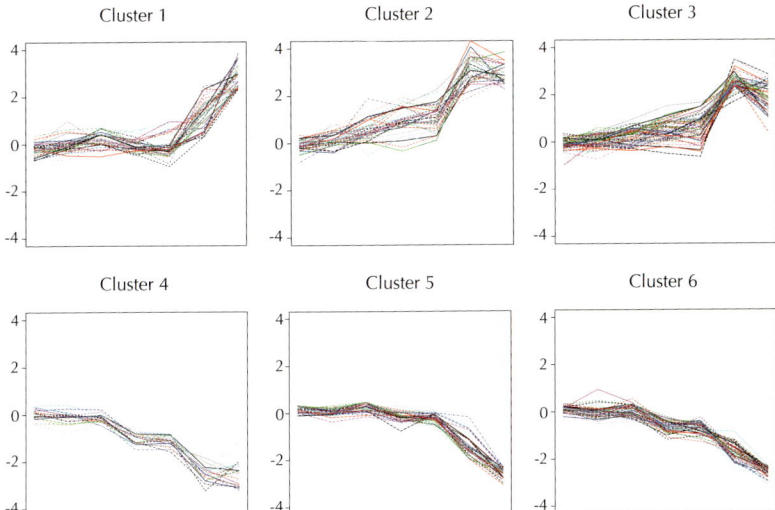

FIGURE 8.14 Applying the Lloyd algorithm (with $k = 6$) to the abridged yeast dataset containing 230 genes results in six clusters revealing six different types of regulatory behavior and containing 37, 36, 58, 19, 36, and 44 genes. Expression vectors for all genes in each of these six clusters are visualized as separate plots.

The plots in Figure 8.14 reveal six patterns of behavior of genes involved in the diauxic shift and raise questions for further biological studies beyond the focus of this chapter. For example, what regulatory mechanisms force the genes in the first cluster to increase their expression? What mechanisms cause the genes in the fourth cluster to decrease their expression? And how do these changes contribute to the diauxic shift?

CHAPTER 8

EXERCISE BREAK: The clustering of the entire 6,400-gene yeast dataset into six clusters implies a clustering of the abridged set containing 230 genes. How does this implied clustering compare to the clustering in Figure 8.14?

Limitations of *k*-Means Clustering

After seeing the Lloyd algorithm in action, it may seem that clustering is easy. If you think so, consider the following question.

STOP and Think: How would you cluster the points in Figure 8.15?

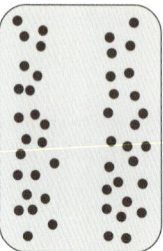 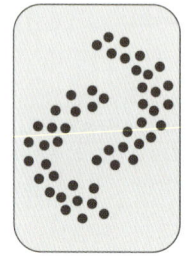 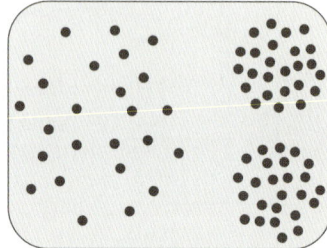

FIGURE 8.15 Difficult clustering problems for $k = 2$ (left and middle) and $k = 3$ (right).

In the case of challenging clustering problems, the Lloyd algorithm sometimes fails to identify what may seem like obvious clusters (Figure 8.16).

STOP and Think: The Lloyd algorithm assigns each data point to its closest center, with ties broken arbitrarily. What are the negative effects of this assignment?

One weakness with our formulation of the *k*-Means Clustering Problem is that it forces us to make a "hard" assignment of each point to only one cluster. This strategy makes little sense for **midpoints**, or points that are approximately equidistant from two centers. To deal with midpoints, our goal is to transition away from a rigid assignment of a data point to a single cluster (Figure 8.17 (left)) and toward a "soft" assignment (Figure 8.17 (right)).

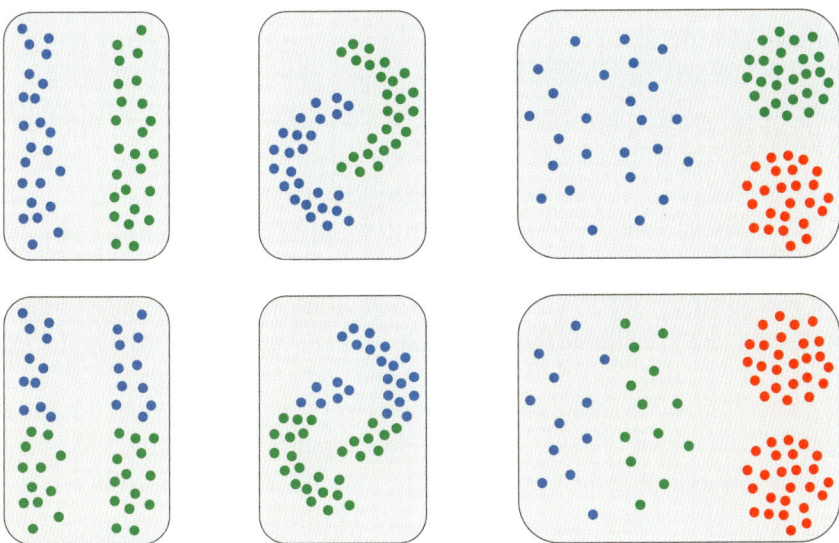

FIGURE 8.16 The human eye (top) and the Lloyd algorithm (bottom) often disagree in the case of elongated clusters (left), clusters with non-globular shapes (middle), and clusters with widely different data point densities (right).

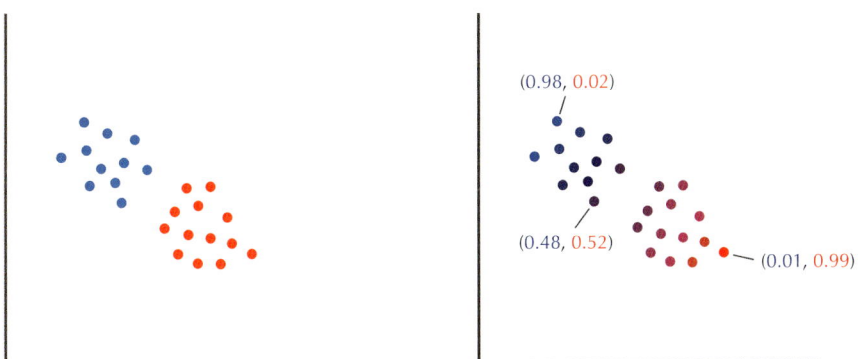

FIGURE 8.17 (Left) The points from Figure 8.6 (left) partitioned into two clusters by the Lloyd algorithm; points are colored red or blue depending on their cluster membership. (Right) We can visualize a soft clustering of the same data into two clusters as assigning each point a pair of numbers representing the point's percentage of "blue" and "red" based on each cluster's "responsibility" for this point. The colors mix to form a color from the red-blue spectrum.

CHAPTER 8

From Coin Flipping to *k*-Means Clustering

Flipping coins with unknown biases

To develop an algorithm for soft clustering, we will introduce a seemingly unrelated analogy. Tom Stoppard's play *Rosencrantz and Guildenstern Are Dead* opens with the title characters flipping a coin over and over, finding that it results in "heads" 157 consecutive times. Rosencrantz and Guildenstern question whether they have become detached from the laws of probability, but if you witnessed such an event, you would probably guess that the coin was biased. Say that your friend flips a biased coin n times, and you would like to estimate the probability θ that a single coin flip results in heads.

> **STOP and Think:** For each value of θ between 0 and 1, you can compute the probability of a given sequence of flips. How would you estimate the value of θ maximizing this probability after watching a series of flips where the coin lands on heads i out of n times?

It seems like the best estimate of θ should be the number of occurrences of heads divided by the total number of coin flips. But how can we prove this? Given a sequence of n coin flips containing i heads, the probability that a coin with bias θ generated this sequence is $f(\theta) = \theta^i \cdot (1-\theta)^{n-i}$. Since the most likely coin bias is the value of θ that maximizes this probability, we will set the derivative of $f(\theta)$ equal to zero,

$$\begin{aligned} f'(\theta) &= i \cdot \theta^{i-1} \cdot (1-\theta)^{n-i} - \theta^i \cdot (n-i) \cdot (1-\theta)^{n-i-1} \\ &= [i \cdot (1-\theta) - \theta \cdot (n-i)] \cdot \theta^{i-1} \cdot (1-\theta)^{n-i-1} \\ &= (i - \theta \cdot n) \cdot \theta^{i-1} \cdot (1-\theta)^{n-i-1} = 0 \,. \end{aligned}$$

Other than $\theta = 0$ and $\theta = 1$, the only solution of this equation is $\theta = i/n$, implying that the observed proportion of heads provides the best estimate for θ.

To make the coin flipping problem a bit more interesting, suppose that your friend secretly switches between two coins A and B that look identical but have unknown biases θ_A and θ_B. After observing a sequence of coin flips, your goal is to estimate θ_A and θ_B, which we collectively denote by *Parameters*.

We will simplify the problem by assuming that every n flips, your friend secretly decides to either keep the same coin or switch coins. Five sequences of $n = 10$ flips are shown in Figure 8.18; we will represent the proportion of heads in each of these sequences as a vector,

$$Data = (Data_1, Data_2, Data_3, Data_4, Data_5) = (0.4, 0.9, 0.8, 0.3, 0.7)\,.$$

										Data
H	T	T	T	H	T	T	H	T	H	0.4
H	H	H	H	T	H	H	H	H	H	0.9
H	T	H	H	H	H	T	H	H	0.8
H	T	T	T	T	T	H	H	T	T	0.3
T	H	H	H	T	H	H	H	T	H	0.7

FIGURE 8.18 Five sequences of ten coin flips result in *Data* = (0.4, 0.9, 0.8, 0.3, 0.7). "H" denotes heads and "T" denotes tails.

If you knew that your friend used coin A in the first and fourth sequences of flips, then you would estimate θ_A by computing the proportion of heads in these sequences,

$$\theta_A = \frac{Data_1 + Data_4}{2} = \frac{0.4 + 0.3}{2} = 0.35.$$

You would then estimate θ_B as the proportion of heads in the remaining three sequences,

$$\theta_B = \frac{Data_2 + Data_3 + Data_5}{3} = \frac{0.9 + 0.8 + 0.7}{3} = 0.8.$$

We will represent this choice of coins as a binary vector *HiddenVector* = (1, 0, 0, 1, 0), where a 1 in the k-th position denotes that coin A was used to generate the k-th sequence of flips, and a 0 denotes that coin B was used. This notation allows us to rewrite the equations for *Parameters* in terms of *Data* and *HiddenVector*:

$$\theta_A = \frac{\sum_i HiddenVector_i \cdot Data_i}{\sum_i HiddenVector_i} = \frac{1 \cdot 0.4 + 0 \cdot 0.9 + 0 \cdot 0.8 + 1 \cdot 0.3 + 0 \cdot 0.7}{1 + 0 + 0 + 1 + 0} = 0.35$$

$$\theta_B = \frac{\sum_i (1 - HiddenVector_i) \cdot Data_i}{\sum_i (1 - HiddenVector_i)} = \frac{0 \cdot 0.4 + 1 \cdot 0.9 + 1 \cdot 0.8 + 0 \cdot 0.3 + 1 \cdot 0.7}{0 + 1 + 1 + 0 + 1} = 0.80$$

where i runs over all data points.

The expression $\sum_i HiddenVector_i \cdot Data_i$ is the **dot product** of vectors *HiddenVector* and *Data*, written *HiddenVector · Data*. Define the **all ones vector**, written $\vec{1}$, as the vector consisting of all ones and whose length is equal to that of *HiddenVector*. This allows us to write $\sum_i HiddenVector_i$ as the dot product $HiddenVector \cdot \vec{1}$ and $\sum_i (1 - HiddenVector_i)$ as the dot product $(\vec{1} - HiddenVector) \cdot \vec{1}$. As a result, the above equations become

$$\theta_A = \frac{HiddenVector \cdot Data}{HiddenVector \cdot \vec{1}}$$

$$\theta_B = \frac{(\vec{1} - HiddenVector) \cdot Data}{(\vec{1} - HiddenVector) \cdot \vec{1}}$$

CHAPTER 8

STOP and Think: We just saw that given *Data* and *HiddenVector*, we can find *Parameters* = (θ_A, θ_B). If you are given *Data* and *Parameters*, can you find the most likely choice of *HiddenVector*?

If we know *Parameters*, then deciding on the most likely choice of *HiddenVector* corresponds to determining whether coin A or coin B was more likely to have generated the n observed flips in each of the five coin flipping sequences. For example, suppose we know that *Parameters* = $(\theta_A, \theta_B) = (0.6, 0.82)$. If coin A was used to generate the fifth sequence of flips, then the probability that it generated the outcome in Figure 8.18 is

$$\theta_A^7 (1 - \theta_A)^3 = 0.6^7 \cdot 0.4^3 \approx 0.00179.$$

If coin B was used to generate the fifth sequence, then the probability that it generated this outcome is

$$\theta_B^7 (1 - \theta_B)^3 = 0.82^7 \cdot 0.18^3 \approx 0.00145.$$

Since $0.00179 > 0.00145$, we would set *HiddenVector*$_5$ equal to 1.

EXERCISE BREAK: Determine the rest of the entries in *HiddenVector* for *Parameters* = (0.6, 0.82) and the sequences of coin flips in Figure 8.18.

More generally, let $\Pr(Data_i | \theta)$ denote the **conditional probability** of generating the outcome $Data_i$ given a coin with bias θ,

$$\Pr(Data_i | \theta) = \theta^{n \cdot Data_i} (1 - \theta)^{n \cdot (1 - Data_i)}.$$

If $\Pr(Data_i | \theta_A) > \Pr(Data_i | \theta_B)$, then coin A is more likely to have generated the i-th sequence of flips, and we set *HiddenVector*$_i$ equal to 1. If $\Pr(Data_i | \theta_A) < \Pr(Data_i | \theta_B)$, then coin B is more likely, and we set *HiddenVector*$_i$ equal to 0. Ties are broken arbitrarily.

In summary, if *HiddenVector* is known and *Parameters* is unknown, then we can reconstruct the most likely *Parameters* = (θ_A, θ_B):

$$(Data, HiddenVector, ?) \rightarrow Parameters$$

Likewise, if *Parameters* is known and *HiddenVector* is unknown, then we can reconstruct the most likely *HiddenVector*:

$$(Data, ?, Parameters) \rightarrow HiddenVector$$

Our original problem, however, was that both *HiddenVector* and *Parameters* are unknown:

$$(Data, ?, ?) \rightarrow ???$$

Where is the computational problem?

You may have noticed that we have not formulated the computational problem that we are trying to solve. So define the conditional probability of generating a sequence of coin flips *Data$_i$* given *HiddenVector* and *Parameters* as

$$\Pr(Data_i | HiddenVector, Parameters) = \begin{cases} \Pr(Data_i | \theta_A) & \text{if } HiddenVector_i = 1 \\ \Pr(Data_i | \theta_B) & \text{if } HiddenVector_i = 0 \end{cases}$$

Furthermore, define the conditional probability of generating *Data* given *HiddenVector* and *Parameters* as

$$\Pr(Data | HiddenVector, Parameters) = \prod_{i=1}^{n} \Pr(Data_i | HiddenVector, Parameters).$$

Given *Data*, the computational problem we are trying to solve is to find *HiddenVector* and *Parameters* maximizing $\Pr(Data | HiddenVector, Parameters)$.

From coin flipping to the Lloyd algorithm

Identifying *HiddenVector* and *Parameters* from *Data* may appear hopeless, but we have already learned that starting from a random guess is not necessarily a bad idea. We will therefore start from an arbitrary choice of *Parameters* $= (\theta_A, \theta_B)$ and immediately reconstruct the most likely *HiddenVector*:

$$(Data, ?, Parameters) \rightarrow HiddenVector$$

As soon as we know *HiddenVector*, we will question the wisdom of our initial choice of *Parameters* and re-estimate *Parameters'*:

$$(Data, HiddenVector, ?) \rightarrow Parameters'$$

As illustrated in Figure 8.19 for the initial choice of *Parameters* = (0.6, 0.82), we repeat these two steps and hope that *Parameters* and *HiddenVector* are moving closer to the values that maximize $\Pr(Data | HiddenVector, Parameters)$,

$$\begin{aligned}
(Data, ?, Parameters) &\rightarrow (Data, HiddenVector, Parameters) \\
&\rightarrow (Data, HiddenVector, \quad ? \quad) \\
&\rightarrow (Data, HiddenVector, Parameters') \\
&\rightarrow (Data, \quad ? \quad , Parameters') \\
&\rightarrow (Data, HiddenVector', Parameters') \\
&\rightarrow \cdots
\end{aligned}$$

CHAPTER 8

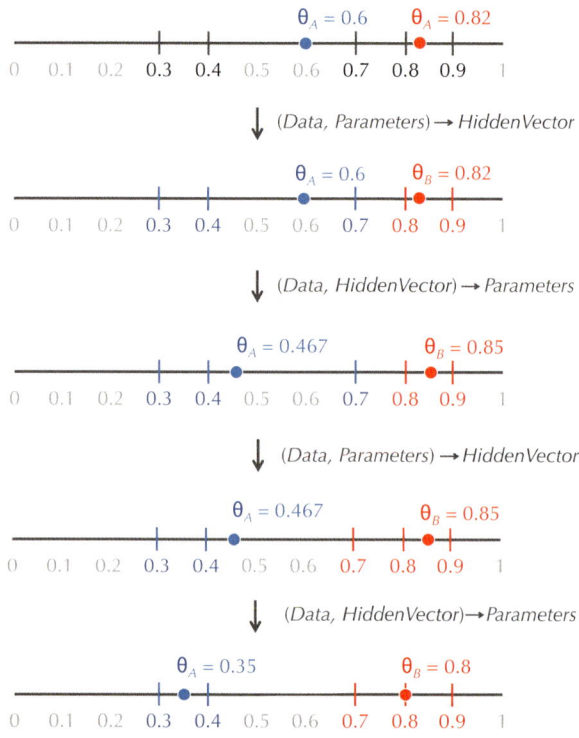

FIGURE 8.19 Starting with *Parameters* = (0.6, 0.82) results in *HiddenVector* = (1, 0, 0, 1, 1) for *Data* = (0.4, 0.9, 0.8, 0.3, 0.7). We then update *Parameters* as (0.467, 0.85), which in turn results in *HiddenVector* = (1, 0, 0, 1, 0). This new vector leads to the assignment of *Parameters* as (0.35, 0.8), at which point the process has terminated because *HiddenVector* will not change in the next step.

EXERCISE BREAK: Prove that this process terminates, i.e., that *HiddenVector* and *Parameters* eventually stop changing between iterations.

STOP and Think: If *HiddenVector* consists of all zeroes, then there are no flips with coin A, and the formula for computing θ_A is invalid. What would you do to address this complication?

Return to clustering

Figure 8.19 has revealed that although our coin flipping analogy seems like a different problem, it is really just a one-dimensional clustering problem in disguise!

STOP and Think: What are *Data*, *HiddenVector*, and *Parameters* in the Lloyd algorithm?

Given n data points in m-dimensional space $Data = (Data_1, \ldots, Data_n)$, we represent their assignment to k clusters as an n-dimensional vector

$$HiddenVector = (HiddenVector_1, \ldots, HiddenVector_n),$$

where each $HiddenVector_i$ can take integer values from 1 to k. We will then represent the k centers as k points in m-dimensional space, $Parameters = (\theta_1, \ldots, \theta_k)$.

In k-means clustering, similarly to the coin flipping analogy, we are given *Data*, but *HiddenVector* and *Parameters* are unknown. The Lloyd algorithm starts from randomly chosen *Parameters*, and we can now rewrite its two main steps as follows:

- Centers to Clusters: $(Data, ?, Parameters) \rightarrow HiddenVector$

- Clusters to Centers: $(Data, HiddenVector, ?) \rightarrow Parameters$

The only difference between the coin flipping algorithm and the Lloyd algorithm for k-means clustering is how they execute the "Centers to Clusters" step. In the former, we compute $HiddenVector_i$ by comparing $\Pr(Data_i|\theta_A)$ with $\Pr(Data_i|\theta_B)$, whereas in the latter, we assign a point to the cluster containing the center nearest to that point.

STOP and Think: Consider the following questions regarding coin flipping and clustering.

- Is it fair to always select coin A if $\Pr(Data_i|\theta_A)$ is only slightly larger than $\Pr(Data_i|\theta_B)$?

- Is it fair to always assign a point to a center if this center is only slightly closer to the point than another center?

Making Soft Decisions in Coin Flipping

Expectation maximization: the E-step

We will now use our coin flipping analogy to motivate a soft version of k-means clustering. Given $Parameters = (\theta_A, \theta_B)$, we can make hard decisions for $HiddenVector$ by comparing $\Pr(Data_i|\theta_A)$ with $\Pr(Data_i|\theta_B)$. But this does not mean that we are

CHAPTER 8

certain which coin was used. If $\Pr(Data_i|\theta_B)$ were approximately equal to $\Pr(Data_i|\theta_A)$, then our confidence that coin B was used would be approximately 50%. On the other hand, if $\Pr(Data_i|\theta_B)$ were much larger than $\Pr(Data_i|\theta_A)$, then we would be almost positive that coin B was used. More generally, we can speak of our confidence that a coin was used as the "responsibility" of this coin for a given sequence of flips. (The responsibilities should sum to 1.)

In terms of k-means clustering, if a data point is a midpoint between two centers, then each of these centers should have about the same responsibility for attracting it to their clusters. As with coin flipping, the responsibilities of all centers for a given data point should sum to 1.

STOP and Think: Given *Parameters* = (θ_A, θ_B) = (0.6, 0.82) and the sequence of coin flips "THHHTHHHTH", how would you compute responsibilities for coins A and B?

To answer the preceding question, we have seen that $\Pr(0.7|\theta_A) = 0.6^7 \cdot 0.4^3 \approx 0.00179$ and that $\Pr(0.7|\theta_B) = 0.82^7 \cdot 0.18^3 \approx 0.00145$. Before, we rigidly concluded that coin A was more likely. Now, since coin A is more likely to generate seven heads in a sequence of ten coin flips, we should assign a larger responsibility to coin A than to coin B. One possible way to assign these responsibilities is given by the formulas

$$\frac{\Pr(0.7|\theta_A)}{\Pr(0.7|\theta_A) + \Pr(0.7|\theta_B)} = \frac{0.00179}{0.00179 + 0.00145} \approx 0.55$$

$$\frac{\Pr(0.7|\theta_B)}{\Pr(0.7|\theta_A) + \Pr(0.7|\theta_B)} = \frac{0.00145}{0.00179 + 0.00145} \approx 0.45$$

As a result, instead of a vector *HiddenVector*, we now have a 2×5 **responsibility profile** *HiddenMatrix* that can be constructed from *Data* and *Parameters*,

$$(Data, ?, Parameters) \to HiddenMatrix.$$

We call this transition the **E-step** (Figure 8.20).

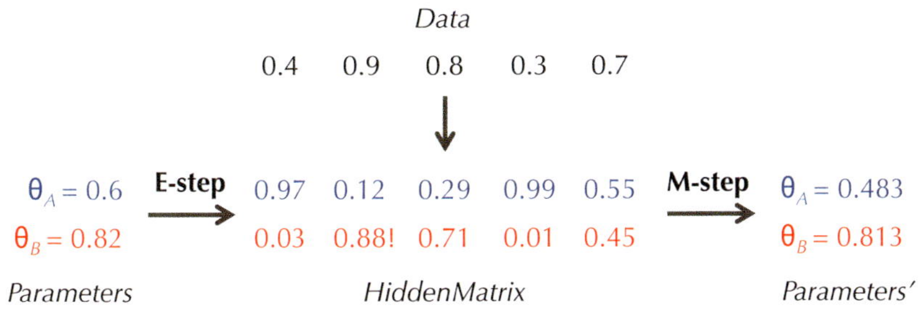

	Data						
	0.4	0.9	0.8	0.3	0.7		
			↓				
θ_A = 0.6 **E-step**	0.97	0.12	0.29	0.99	0.55	**M-step**	θ_A = 0.483
θ_B = 0.82 →	0.03	0.88!	0.71	0.01	0.45	→	θ_B = 0.813
Parameters		HiddenMatrix					Parameters'

FIGURE 8.20 In the E-step, we compute *HiddenMatrix* from *Data* = (0.4, 0.9, 0.8, 0.3, 0.7) and *Parameters* = (0.6, 0.82). In the M-step, we compute an updated *Parameters'* from *HiddenMatrix* and *Data*.

Expectation maximization: the M-step

When making hard assignments, we computed *Parameters* from *Data* and *HiddenVector* as follows:

$$\theta_A = \frac{HiddenVector \cdot Data}{HiddenVector \cdot \vec{1}}$$

$$\theta_B = \frac{(\vec{1} - HiddenVector) \cdot Data}{(\vec{1} - HiddenVector) \cdot \vec{1}}.$$

To make soft assignments, note that the assignment of outcomes to the two coins can be represented by the binary responsibility matrix below. An occurrence of 1 in the *i*-th position of the first row means that we conclude that coin *A* generated the *i*-th sequence of flips, and an occurrence of 1 in the second row means that we conclude that coin *B* generated the *i*-th sequence of flips:

$$HiddenMatrix \quad \begin{matrix} 0 & 1 & 1 & 0 & 1 \\ 1 & 0 & 0 & 1 & 0 \end{matrix}$$

Thus, the first row of *HiddenMatrix*, denoted $HiddenMatrix_A$, is just *HiddenVector*, and the second row of *HiddenMatrix*, denoted $HiddenMatrix_B$, is just $\vec{1}$ − *HiddenVector*. We can therefore rewrite the previous formulas for θ_A and θ_B in terms of *HiddenMatrix*:

$$\theta_A = \frac{HiddenMatrix_A \cdot Data}{HiddenMatrix_A \cdot \vec{1}}$$

$$\theta_B = \frac{HiddenMatrix_B \cdot Data}{HiddenMatrix_B \cdot \vec{1}}$$

CHAPTER 8

For the responsibility matrix in Figure 8.20, we can now recompute *Parameters* as follows:

$$\theta_A = \frac{0.97 \cdot 0.4 + 0.12 \cdot 0.9 + 0.29 \cdot 0.8 + 0.99 \cdot 0.3 + 0.55 \cdot 0.7}{0.97 + 0.12 + 0.29 + 0.99 + 0.55} = \frac{1.41}{2.92} \approx 0.483$$

$$\theta_B = \frac{0.03 \cdot 0.4 + 0.88 \cdot 0.9 + 0.71 \cdot 0.8 + 0.01 \cdot 0.3 + 0.45 \cdot 0.7}{0.03 + 0.88 + 0.71 + 0.01 + 0.45} = \frac{1.69}{2.08} \approx 0.813$$

STOP and Think: Notice that the soft parameter choices $\theta_A = 0.483$ and $\theta_B = 0.813$ are a little closer to each other than the hard parameter choices $\theta_A = 0.467$ and $\theta_B = 0.85$. Why do you think that this is the case?

In general, the transition

$$(Data, HiddenMatrix, ?) \rightarrow Parameters$$

is called the **M-step**.

The expectation maximization algorithm

The **expectation maximization algorithm** starts with a random choice of *Parameters*. It then alternates between the E-step, in which we compute a responsibility matrix *HiddenMatrix* for *Data* given *Parameters*:

$$(Data, ?, Parameters) \rightarrow HiddenMatrix$$

and the M-step, in which we re-estimate *Parameters* using *HiddenMatrix*:

$$(Data, HiddenMatrix, ?) \rightarrow Parameters$$

EXERCISE BREAK: Carry out a few more steps of the expectation maximization algorithm for the data in Figure 8.20. When should we stop the algorithm?

Soft *k*-Means Clustering

Applying expectation maximization to clustering

We are now ready to use the expectation maximization algorithm to modify the Lloyd algorithm into a **soft *k*-means clustering algorithm**. This algorithm starts from randomly chosen centers and iterates the following two steps:

- **Centers to Soft Clusters (E-step):** After centers have been selected, assign each data point a "responsibility" for each cluster, where higher responsibilities correspond to stronger cluster membership.

- **Soft Clusters to Centers (M-step):** After data points have been assigned to soft clusters, compute new centers.

Centers to soft clusters

We begin with the "Centers to Soft Clusters" step. We have already used the term "center of gravity" when computing centers; if we think about the centers as stars and the data points as planets, then the closer a point is to a center, the stronger that center's "pull" should be on the point. Given k centers $Centers = (x_1, \ldots x_k)$ and n points $Data = (Data_1, \ldots, Data_n)$, we therefore need to construct a $k \times n$ responsibility matrix $HiddenMatrix$ for which $HiddenMatrix_{i,j}$ is the pull of center i on data point j. This pull can be computed according to the Newtonian inverse-square law of gravitation,

$$HiddenMatrix_{i,j} = \frac{1/d(Data_j, x_i)^2}{\sum_{\text{all centers } x_i} 1/d(Data_j, x_i)^2}.$$

Unfortunately for Newton fans, the following **partition function** from statistical physics often works better in practice:

$$HiddenMatrix_{i,j} = \frac{e^{-\beta \cdot d(Data_j, x_i)}}{\sum_{\text{all centers } x_i} e^{-\beta \cdot d(Data_j, x_i)}}$$

In this formula, e is the base of the natural logarithm ($e \approx 2.718$), and β is a parameter reflecting the amount of flexibility in our soft assignment and called — appropriately enough — the **stiffness parameter**. Figure 8.21 illustrates different approaches for computing $HiddenMatrix$ when $Data$ represents points in one-dimensional space.

STOP and Think: How does the assignment of the points in Figure 8.21 to soft clusters change as $\beta \to \infty$ or as $\beta \to -\infty$? What about the case that $\beta = 0$?

EXERCISE BREAK: Compute $HiddenMatrix$ using the Newtonian inverse-square law for the three centers and eight data points shown in Figure 8.10 (left).

CHAPTER 8

```
           0.992  0.988  0.500  0.012  0.008
           0.008  0.012  0.500  0.988  0.992

           0.924  0.881  0.500  0.119  0.076
           0.076  0.119  0.500  0.881  0.924

           0.993  0.982  0.500  0.118  0.007
           0.007  0.018  0.500  0.982  0.993
```

FIGURE 8.21 (Top) Five one-dimensional points *Data* = (-3, -2, 0, +2, +3) with two centers (shown in blue and red) *Centers* = {-2.5, +2.5}. (Bottom) Three versions of *HiddenMatrix* constructed for *Data* and *Centers*, using the Newtonian inverse-square law (first matrix) and the partition function with stiffness $\beta = 0.5$ (second matrix), and $\beta = 1$ (third matrix).

Soft clusters to centers

When we implemented the M-step for coin flipping, we obtained the following formulas for θ_A and θ_B:

$$\theta_A = \frac{HiddenMatrix_A \cdot Data}{HiddenMatrix_A \cdot \vec{1}}$$

$$\theta_B = \frac{HiddenMatrix_B \cdot Data}{HiddenMatrix_B \cdot \vec{1}}$$

In soft *k*-means clustering, if we let $HiddenMatrix_i$ denote the *i*-th row of *HiddenMatrix*, then we can update center x_i using an analogue of the above formulas. Specifically, we will define the *j*-th coordinate of center x_i, denoted $x_{i,j}$, as

$$x_{i,j} = \frac{HiddenMatrix_i \cdot Data^j}{HiddenMatrix_i \cdot \vec{1}}$$

Here, $Data^j$ is the *n*-dimensional vector holding the *j*-th coordinates of the *n* points in *Data*. The updated center x_i is called a **weighted center of gravity** of the points *Data*.

Computing weighted centers of gravity for the *HiddenMatrix* at the bottom of Figure 8.21 produces the following updated centers:

$$x_1 = \frac{0.993 \cdot (-3) + 0.982 \cdot (-2) + 0.500 \cdot (0) + 0.018 \cdot (2) + 0.007 \cdot (3)}{0.993 + 0.982 + 0.500 + 0.018 + 0.007} = -1.955$$

$$x_2 = \frac{0.007 \cdot (-3) + 0.018 \cdot (-2) + 0.500 \cdot (0) + 0.982 \cdot (2) + 0.993 \cdot (3)}{0.007 + 0.018 + 0.500 + 0.982 + 0.993} = 1.955$$

You are now ready to implement the expectation maximization algorithm for soft k-means clustering.

STOP and Think: Does the soft k-means clustering algorithm terminate? If not, how would you modify it to ensure that it does not run forever?

EXERCISE BREAK: Recompute centers using the data points from the exercise on page 100 along with the responsibility matrix obtained as the result of this algorithm.

EXERCISE BREAK: Apply soft k-means clustering to the abridged yeast diauxic shift gene expression data, and compare the results with those of the Lloyd algorithm.

Hierarchical Clustering

Introduction to distance-based clustering

In Chapter 7, we discussed two approaches to evolutionary tree reconstruction with different strengths and weaknesses: distance-based algorithms (including the neighbor-joining algorithm), and alignment-based algorithms (including the algorithm for the Small Parsimony Problem). Similarly, biologists do not always analyze the $n \times m$ gene expression matrix directly. Instead, they sometimes first transform this matrix into an $n \times n$ **distance matrix** D, where $D_{i,j}$ indicates the *distance* between the expression vectors for genes i and j (Figure 8.22 (top right)). In this section, we will see how to use a distance matrix to partition genes into clusters (see **DETOUR: Transforming an Expression Matrix into a Distance/Similarity Matrix** for more details).

In previous sections, we assumed that we were working with a fixed number of clusters k. But in practice, clusters often have subclusters, which have subsubclusters,

CHAPTER 8

	1 hr	2 hr	3 hr
g_1	10.0	8.0	10.0
g_2	10.0	0.0	9.0
g_3	4.0	8.5	3.0
g_4	9.5	0.5	8.5
g_5	4.5	8.5	2.5
g_6	10.5	9.0	12.0
g_7	5.0	8.5	11.0
g_8	3.7	8.7	2.0
g_9	9.7	2.0	9.0
g_{10}	10.2	1.0	9.2

	g_1	g_2	g_3	g_4	g_5	g_6	g_7	g_8	g_9	g_{10}
g_1	0.0	8.1	9.2	7.7	9.3	2.3	5.1	10.2	6.1	7.0
g_2	8.1	0.0	12.0	0.9	12.0	9.5	10.1	12.8	2.0	1.0
g_3	9.2	12.0	0.0	11.2	0.7	11.1	8.1	1.1	10.5	11.5
g_4	7.7	0.9	11.2	0.0	11.2	9.2	9.5	12.0	1.6	1.1
g_5	9.3	12.0	0.7	11.2	0.0	11.2	8.5	1.0	10.6	11.6
g_6	2.3	9.5	11.1	9.2	11.2	0.0	5.6	12.1	7.7	8.5
g_7	5.1	10.1	8.1	9.5	8.5	5.6	0.0	9.1	8.3	9.3
g_8	10.2	12.8	1.1	12.0	1.0	12.1	9.1	0.0	11.4	12.4
g_9	6.1	2.0	10.5	1.6	10.6	7.7	8.3	11.4	0.0	1.1
g_{10}	7.0	1.0	11.5	1.1	11.6	8.5	9.3	12.4	1.1	0.0

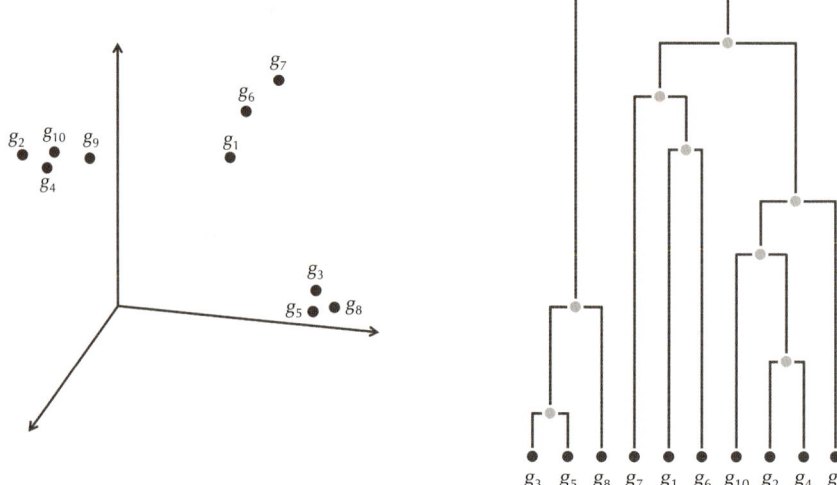

FIGURE 8.22 A toy gene expression matrix of ten genes measured at three time points (top left), the distance matrix based on Euclidean distance (top right), the gene expression vectors as points in three-dimensional space (bottom left), and the tree produced from the distance matrix by the hierarchical clustering algorithm (bottom right). Leaves correspond to genes; internal nodes correspond to clusters of genes.

and so on. To capture this cluster stratification, the **hierarchical clustering** algorithm uses an $n \times n$ distance matrix D to organize n data points into a tree (Figure 8.22 (bottom right)). As shown in Figure 8.23, a horizontal line crossing the tree in i places divides the n genes into i clusters.

 EXERCISE BREAK: Figure 8.23 illustrates two ways of clustering the data from Figure 8.22 using a tree. Find the remaining eight ways of clustering these data using the same tree.

HOW DID YEAST BECOME A WINE MAKER?

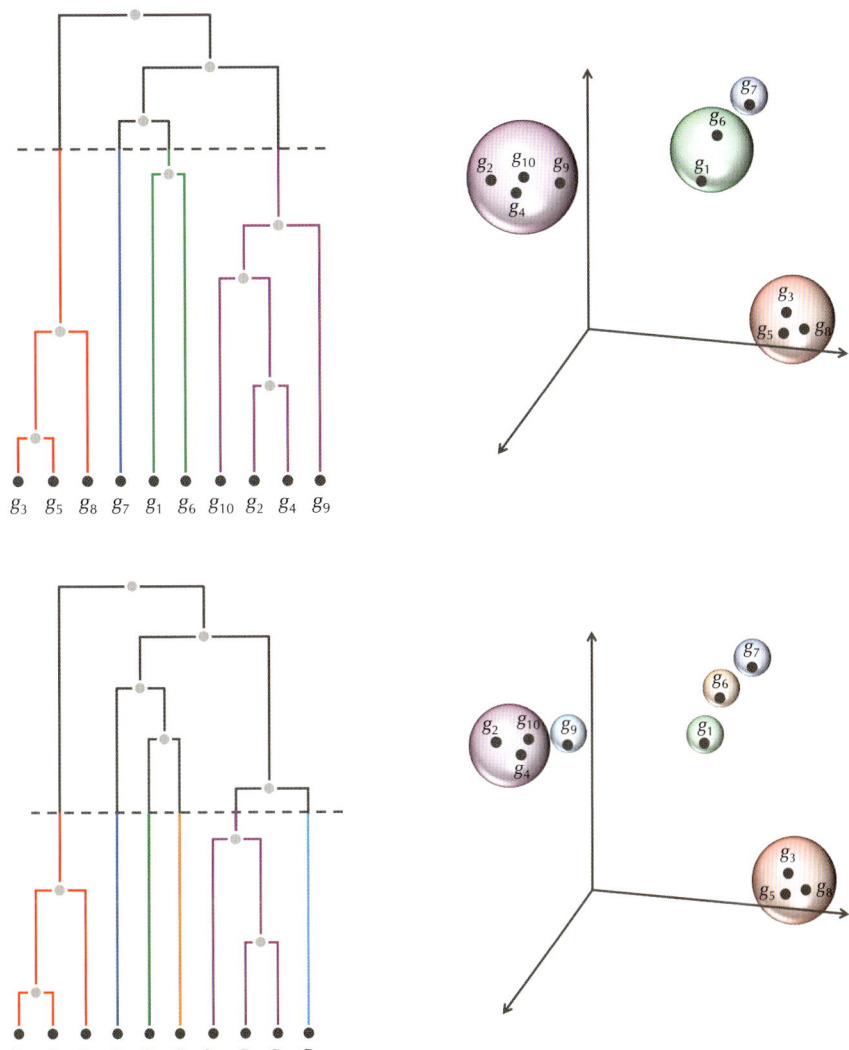

FIGURE 8.23 A tree with *n* leaves imposes *n* different ways of partitioning the data into clusters. (Top) The horizontal line through the tree (left) crosses the data in four places and partitions the data into four clusters (right). (Bottom) The same tree with a different horizontal line (left) partitions the data into six clusters (right).

CHAPTER 8

Inferring clusters from a tree

HIERARCHICALCLUSTERING, whose pseudocode is shown below, progressively generates n different partitions of the underlying data into clusters, all represented by a tree in which each node is labeled by a cluster of genes. The first partition has n single-element clusters represented by the leaves of the tree, with each element forming its own cluster. The second partition merges the two "closest" clusters into a single cluster consisting of two elements. In general, the i-th partition merges the two closest clusters from the $(i-1)$-th partition and has $n - i + 1$ clusters. We hope this algorithm looks familiar — it is **UPGMA** (from Chapter 7) in disguise.

HIERARCHICALCLUSTERING(*D*, *n*)
    Clusters ← *n* single-element clusters labeled $1, \ldots, n$
    construct a graph *T* with *n* isolated nodes labeled by single elements $1, \ldots, n$
    while there is more than one cluster
        find the two closest clusters C_i and C_j (break ties arbitrarily)
        merge C_i and C_j into a new cluster C_{new} with $|C_i| + |C_j|$ elements
        add a new node labeled by cluster C_{new} to *T*
        connect node C_{new} to C_i and C_j by directed edges
        remove the rows and columns of *D* corresponding to C_i and C_j
        remove C_i and C_j from *Clusters*
        add a row/column to *D* for C_{new} by computing $D(C_{new}, C)$ for each *C* in *Clusters*
        add C_{new} to *Clusters*
    root ← the node in *T* corresponding to the remaining cluster
    return *T*

Note that we have not yet defined how **HIERARCHICALCLUSTERING** computes the distance $D(C_{new}, C)$ between a newly formed cluster C_{new} and each old cluster C. In practice, clustering algorithms vary in how they compute these distances, with results that can vary greatly. One commonly used approach (Figure 8.24) defines the distance between clusters C_1 and C_2 as the smallest distance between any pair of elements from these clusters,

$$D_{\min}(C_1, C_2) = \min_{\text{all points } i \text{ in cluster } C_1, \text{ all points } j \text{ in cluster } C_2} D_{i,j}.$$

The distance function that we encountered with **UPGMA** uses the average distance between elements in two clusters,

HOW DID YEAST BECOME A WINE MAKER?

$$D_{\text{avg}}(C_1, C_2) = \frac{\sum_{\text{all points } i \text{ in cluster } C_1} \sum_{\text{all points } j \text{ in cluster } C_2} D_{i,j}}{|C_1| \cdot |C_2|}.$$

	g_1	g_2	g_3	g_4	g_5	g_6	g_7	g_8	g_9	g_{10}
g_1	0.0	8.1	9.2	7.7	9.3	2.3	5.1	10.2	6.1	7.0
g_2	8.1	0.0	12.0	0.9	12.0	9.5	10.1	12.8	2.0	1.0
g_3	9.2	12.0	0.0	11.2	**0.7**	11.1	8.1	1.1	10.5	11.5
g_4	7.7	0.9	11.2	0.0	11.2	9.2	9.5	12.0	1.6	1.1
g_5	9.3	12.0	0.7	11.2	0.0	11.2	8.5	1.0	10.6	11.6
g_6	2.3	9.5	11.1	9.2	11.2	0.0	5.6	12.1	7.7	8.5
g_7	5.1	10.1	8.1	9.5	8.5	5.6	0.0	9.1	8.3	9.3
g_8	10.2	12.8	1.1	12.0	1.0	12.1	9.1	0.0	11.4	12.4
g_9	6.1	2.0	10.5	1.6	10.6	7.7	8.3	11.4	0.0	1.1
g_{10}	7.0	1.0	11.5	1.1	11.6	8.5	9.3	12.4	1.1	0.0

	g_1	g_2	g_3, g_5	g_4	g_6	g_7	g_8	g_9	g_{10}
g_1	0.0	8.1	9.2	7.7	2.3	5.1	10.2	6.1	7.0
g_2	8.1	0.0	12.0	**0.9**	9.5	10.1	12.8	2.0	1.0
g_3, g_5	9.2	12.0	0.0	11.2	11.1	8.1	1.0	10.5	11.5
g_4	7.7	0.9	11.2	0.0	9.2	9.5	12.0	1.6	1.1
g_6	2.3	9.5	11.1	9.2	0.0	5.6	12.1	7.7	8.5
g_7	5.1	10.1	8.1	9.5	5.6	0.0	9.1	8.3	9.3
g_8	10.2	12.8	1.0	12.0	12.1	9.1	0.0	11.4	12.4
g_9	6.1	2.0	10.5	1.6	7.7	8.3	11.4	0.0	1.1
g_{10}	7.0	1.0	11.5	1.1	8.5	9.3	12.4	1.1	0.0

	g_1	g_2, g_4	g_3, g_5	g_6	g_7	g_8	g_9	g_{10}
g_1	0.0	7.7	9.2	2.3	5.1	10.2	6.1	7.0
g_2, g_4	7.7	0.0	11.2	9.2	9.5	12.0	1.6	1.0
g_3, g_5	9.2	11.2	0.0	11.1	8.1	**1.0**	10.5	11.5
g_6	2.3	9.2	11.1	0.0	5.6	12.1	7.7	8.5
g_7	5.1	9.5	8.1	5.6	0.0	9.1	8.3	9.3
g_8	10.2	12.0	1.0	12.1	9.1	0.0	11.4	12.4
g_9	6.1	1.6	10.5	7.7	8.3	11.4	0.0	1.1
g_{10}	7.0	1.0	11.5	8.5	9.3	12.4	1.1	0.0

FIGURE 8.24 HIERARCHICALCLUSTERING in action. (Top left) The distance matrix from Figure 8.22 (top left), with its minimum element shown in red, corresponding to genes g_3 and g_5. (Top right) Merging the single-element clusters containing g_3 and g_5. (Middle left) The updated distance matrix after computing $D_{\min}$ for the new cluster with respect to each other (single-element) cluster, with its minimum element shown in red. (Middle right) Merging the two clusters corresponding to the minimum element. (Bottom) Updating the distance matrix (left) and merging two additional clusters (right). Subsequent steps will reconstruct the tree from Figure 8.22.

EXERCISE BREAK: Apply **HIERARCHICALCLUSTERING** to the distance matrix in Figure 8.22 using D_{avg} instead of $D_{\min}$.

CHAPTER 8

EXERCISE BREAK: Apply **HIERARCHICALCLUSTERING** (with D_{avg}) to the abridged 230-gene yeast dataset, and partition this dataset into six clusters. Do you expect these clusters to be roughly the same as the clusters shown in Figure 8.14? If not, should we be concerned?

Analyzing the diauxic shift with hierarchical clustering

Figure 8.25 visualizes expression vectors for each of the six clusters obtained after applying **HIERARCHICALCLUSTERING** (using D_{avg}) to the yeast dataset.

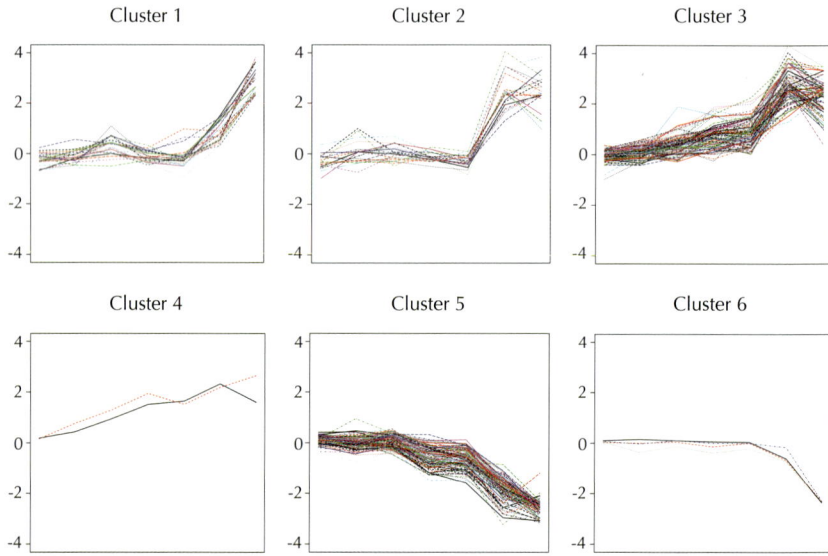

FIGURE 8.25 Applying **HIERARCHICALCLUSTERING** to the yeast dataset results in six clusters with 22, 20, 87, 2, 95, and 4 genes, respectively.

STOP and Think: **HIERARCHICALCLUSTERING** and the Lloyd algorithm (Figure 8.12) have produced different clusters. Should we be concerned?

EXERCISE BREAK: Implement **HIERARCHICALCLUSTERING** (with D_{min} rather than D_{avg}) and apply it to partition the abridged yeast gene expression dataset into six clusters. How does the result differ from Figure 8.25?

Biologists are not discouraged by the fact that different clustering approaches may produce different clusters because applying these clustering algorithms is often just the first step on the road to discovery (see **DETOUR: Clustering and Corrupted Cliques** `PAGE 115` for yet another clustering approach). For this reason, gene expression studies are typically followed by experimental work to confirm that derived clusters make sense biologically. Once clusters have been generated, further research often focuses on specific genes within these clusters.

For example, each of the clusters in Figure 8.25 can be further analyzed to reveal sub-clusters of genes with even more pronounced expression profiles than the genes in the entire cluster. In particular, cluster 1 contains seven genes exhibiting rather slight changes during the first six checkpoints and a surge in gene expression at the final checkpoint. Biologists discovered that six of these seven genes have the **carbon source response element (CSRE) regulatory motif** with consensus sequence `CATTCATCCG` in their upstream regions. Further analysis of the entire yeast genome revealed that only four other yeast genes have this motif in their upstream region, suggesting that it was a good idea to group these six genes into a sub-cluster within cluster 1.

The more important issue, however, is to understand *why* these six genes are related. Yeast prefers using glucose as an energy source compared to other compounds like ethanol, and so in the presence of glucose, the transcription of genes responsible for metabolizing these less tasty compounds is repressed. Researchers have thus concluded that the CSRE motif somehow helps yeast sense the presence of glucose and activates the six genes in question when the organism runs out of glucose, thus serving as an important component of the diauxic shift.

Finally, if you believe that we have exhausted every possible avenue of clustering, take another look at Figure 8.16 (top). Although **HierarchicalClustering** with D_{min} is able to find the clusters on the left and middle, none of the clustering algorithms we have encountered can produce the clusters on the right. Clustering seems like a straightforward problem in part because the human eye is so adept at grouping points into shapes. After all, computer vision researchers are still trying to teach computers to mimic our visual experience of the world, which is the outcome of millions of years of evolution.

Epilogue: Clustering Tumor Samples

As we mentioned earlier, gene expression analysis has a wide variety of applications, including cancer studies. In 1999, Uri Alon analyzed gene expression data for 2,000 genes from 40 colon tumor tissues and compared them with data from colon tissues belonging to 20 healthy individuals. We can represent his data as a 2,000 × 60 gene expression matrix, where the first 40 columns describe tumor samples and the last 20 columns describe normal samples.

Now, suppose you performed a gene expression experiment with a colon sample from a new patient, corresponding to a 61st column in an augmented gene expression matrix. Your goal is to predict whether this patient has a colon tumor. Since the partition of tissues into two clusters (tumor vs. healthy) is known in advance, it may seem that classifying the sample from a new patient is easy. Indeed, since each patient corresponds to a point in 2,000-dimensional space, we can compute the center of gravity of these points for the tumor sample and for the healthy sample. Afterwards, we can simply check which of the two centers of gravity is closer to the new tissue.

Alternatively, we could perform a blind analysis, pretending that we do not already know the classification of samples into cancerous vs. healthy, and analyze the resulting 2,000 × 61 expression matrix to divide the 61 samples into two clusters. If we obtain a cluster consisting predominantly of cancer tissues, this cluster may help us diagnose colon cancer.

CHALLENGE PROBLEM: These approaches may seem straightforward, but it is unlikely that either of them will reliably diagnose the new patient. Why do you think this is the case? Given Alon's 2,000 × 60 gene expression matrix and gene data from a new patient, derive a superior approach to evaluate whether this patient is likely to have a colon tumor.

Detours

Whole genome duplication or a series of duplications?

WGDs are quickly followed by massive gene loss and rearrangements, making it difficult to reconstruct the pre-duplicated genome. Indeed, as we mentioned in the main text, only 13% of genes have duplicates in modern-day *S. cerevisiae*. How, then, can we argue that *S. cerevisiae* has indeed undergone a WGD instead of a sequence of smaller duplications?

In 2004, Manolis Kellis analyzed *K. waltii*, a related yeast species. By aligning synteny blocks from *K. waltii* and *S. cerevisiae*, he discovered that nearly every synteny block of *K. waltii* aligns to *two* regions of *S. cerevisiae*. Because very few genes in the duplicated *S. cerevisiae* blocks occurred in both blocks (Figure 8.26), Kellis argued that there was indeed a WGD during yeast evolution.

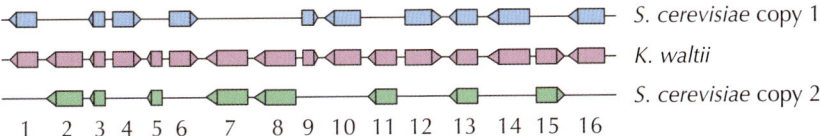

FIGURE 8.26 Two synteny blocks from *S. cerevisiae* (whose respective genes are shown in blue and green) aligned against a synteny block of *K. waltii* (whose genes are shown in purple). Although only three of the sixteen genes in this synteny block have two copies in *S. cerevisiae*, every gene in the *K. waltii* synteny block has a copy in at least one of the two synteny blocks in *S. cerevisiae*. In fact, most synteny blocks in *K. waltii* display this phenomenon, thus suggesting that there was indeed a WGD in *S. cerevisiae*.

STOP and Think: Although Manolis Kellis argued in 2004 that Figure 8.26 provides evidence for a WGD, three years later, Gustavo Caetano-Anollés raised doubts about Kellis's conclusion. Can you devise an alternative explanation for Figure 8.26 and propose a different evolutionary scenario that does not require a WGD?

Measuring gene expression

In the main text, we mentioned that we have encountered three technologies that could be used to measure gene expression. First, in a mass spectrometry experiment (Chapter 4), biologists generate a set of spectra and match them against a proteome.

CHAPTER 8

The number of spectra matching peptides from a given protein offers a proxy for the expression level of this protein. To estimate the protein expression from this proxy, bioinformaticians must account for the varying length of proteins, poor fragmentation of some peptides (leading to difficulties in their identification), and other practical concerns.

Second, in an **RNA sequencing** experiment, we generate reads from a **transcriptome**, or all RNA transcripts present in a cell. By estimating the quantity of each protein-coding RNA transcript in a sample, we obtain a proxy for the expression of the resulting protein. A number of processes affect protein production in the cell in addition to transcription, such as translation, post-translational modifications, and protein degradation. These additional factors can muddle the correlation between the quantity of a transcript and the expression of its corresponding protein.

Third, we can use DNA arrays (Chapter 2) carrying probes (k-mers) aimed at each gene in a species of interest. Each probe is characterized by an intensity, which offers a proxy for the number of transcripts of a given gene present in a sample. A deficiency of DNA arrays is that they only target the identification of known transcripts and often fail to evaluate unknown transcripts. For example, many cancers are caused by rare mutations and would go undetected when using DNA arrays. As a result, RNA sequencing is often more attractive in cancer studies, and it is now the dominant technology for analyzing gene expression.

Microarrays

The microarrays that DeRisi used to study the diauxic shift were manufactured as follows. After capturing many RNA transcripts expressed in yeast cells, DeRisi converted each RNA transcript to **complementary DNA (cDNA)** using an enzyme called **reverse transcriptase**, and spotted these cDNAs on a glass slide. He then hybridized the cDNA against fluorescently labeled RNA from a sample of interest in order to measure the expression levels of various yeast genes.

The amount of cDNA printed on each spot of the microarray can vary greatly, a complication that DeRisi needed to address in order to ensure that fluorescence intensities could be compared across spots and across arrays. He therefore hybridized two samples corresponding to two different timestamps to each array (Figure 8.27). He then labeled the samples with different colors of fluorescent dyes so that the samples could be distinguished by image-processing software.

The expression values obtained from a microarray are represented as the ratio of fluorescent intensities of the two samples. Thus, expression is measured as relative changes in the expression of individual genes between samples and time points. For example, if a gene's expression value is 2, then this gene's expression is twice as large in the first sample; if the expression value is 1/3, then the expression is three times as large in the second sample. Following DeRisi, researchers commonly take the logarithm of these expression ratios to generate expression matrices.

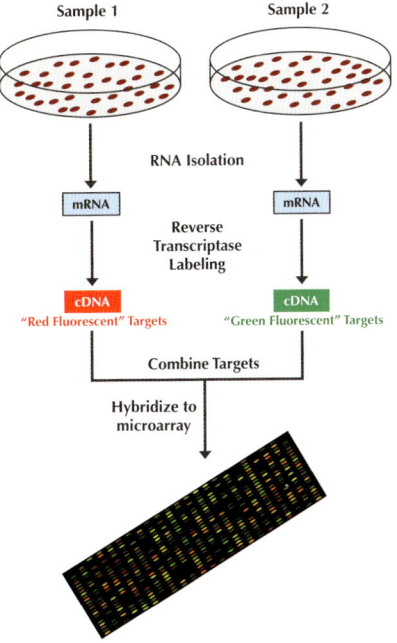

FIGURE 8.27 A microarray against which two samples of fluorescently labeled RNA (red and green) has been hybridized.

Proof of the Center of Gravity Theorem

Note that when $k = 1$, the k-Means Clustering Problem is equivalent to finding a single center point x that minimizes the sum of squared distances from x to points in *Data*.

Our goal is to show that the center of gravity of a set of points *Data* is the unique point that minimizes DISTORTION(*Data*, x) over all possible centers x. Since the squared Euclidean distance between $DataPoint = (DataPoint_1, \ldots, DataPoint_m)$ and center $x =$

CHAPTER 8

$(x_1, \ldots, x_m)$ is equal to $\sum_{1 \leq j \leq m}(DataPoint_j - x_j)^2$, we have that

$$\text{DISTORTION}(Data, x) = \frac{1}{n} \sum_{\text{all points } DataPoint \text{ in } Data} d(DataPoint, x)^2$$

$$= \frac{1}{n} \sum_{\text{all points } DataPoint \text{ in } Data} \sum_{j=1}^{m} (DataPoint_j - x_j)^2$$

$$= \frac{1}{n} \sum_{j=1}^{m} \sum_{\text{all points } DataPoint \text{ in } Data} (DataPoint_j - x_j)^2.$$

The last line of this formula implies that we can independently minimize $\text{DISTORTION}(Data, x)$ in each of m dimensions by minimizing each of the m expressions

$$\sum_{\text{all points } DataPoint \text{ in } Data} (DataPoint_j - x_j)^2.$$

Each of these expressions is a concave-up quadratic function of a single variable x_j. Thus, we can find the minimum of this function by finding where its derivative is equal to zero:

$$\sum_{\text{all points } DataPoint \text{ in } Data} -2 \cdot (DataPoint_j - x_j) = 0.$$

The only solution of this equation is given by

$$x_j = \frac{1}{n} \sum_{\text{all points } DataPoint \text{ in } Data} DataPoint_j,$$

implying that the center's j-th coordinate is the mean value of the j-th coordinates of the data points. In other words, the unique solution of the k-Means Clustering Problem for $k = 1$ is simply the center of gravity of all data points. $\square$

Transforming an expression matrix into a distance/similarity matrix

There are many ways to quantify the similarity between expression vectors $x = (x_1, \ldots, x_m)$ and $y = (y_1, \ldots, y_m)$. One possibility is the dot product, $\sum_{i=1,m} x_i \cdot y_i$. Another is the **Pearson correlation coefficient** $\text{PEARSONCORRELATION}(x, y)$, where

$$\text{PEARSONCORRELATION}(x, y) = \frac{\sum_{i=1}^{m}(x_i - \mu(x)) \cdot (y_i - \mu(y))}{\sqrt{\sum_{i=1}^{m}(x_i - \mu(x))^2 \cdot \sum_{i=1}^{m}(y_i - \mu(y))^2}}.$$

In the above formula, $\mu(x)$ denotes the means of all coordinates of vector x.

STOP and Think: Given a vector x, which vectors y maximize and minimize PEARSONCORRELATION(x,y)?

The Pearson correlation coefficient varies between -1 and 1, where -1 indicates total negative correlation, 0 indicates no correlation, and 1 indicates total positive correlation. Based on the Pearson correlation coefficient, we can define the **Pearson distance** between vectors x and y as

$$\text{PEARSONDISTANCE}(x,y) = 1 - \text{PEARSONCORRELATION}(x,y).$$

EXERCISE BREAK: Compute the Pearson correlation coefficient for the following pairs of vectors:

1. $(\cos\alpha, \sin\alpha)$ and $(\sin\alpha, -\cos\alpha)$ for an arbitrary value of α;
2. $(\sqrt{0.75}, 0.5)$ and $(-\sqrt{0.75}, 0.5)$.

Clustering and corrupted cliques

In expression analysis studies, a similarity matrix R is often transformed into a **similarity graph** $G(R,\theta)$. The nodes of this graph represent genes, and an edge connects genes i and j if and only if the similarity between them ($R_{i,j}$) exceeds a threshold value θ.

STOP and Think: Consider a clustering of genes that satisfies the Good Clustering Principle: the similarity between any two genes within the same cluster exceeds θ, and the similarity between any two genes in different clusters is less than θ. What does the similarity graph $G(R,\theta)$ look like for these genes?

If clusters satisfy the Good Clustering Principle, then there should be some value of θ such that each connected component of $G(R,\theta)$ is a **clique**, or a graph in which every pair of nodes are connected by an edge (Figure 8.28). In general, a graph whose connected components are all cliques is called a **clique graph**.

Errors in expression data and the absence of a universal threshold θ often result in corrupted similarity graphs whose connected components are not cliques (Figure 8.29). Either genes from the same cluster may have a similarity value falling below θ, thus removing edges from a clique, or genes from different clusters may have a similarity value exceeding θ, thus adding edges between different cliques. This observation leads us to ask how to transform a corrupted similarity graph into a clique graph using the smallest number of edge additions and deletions.

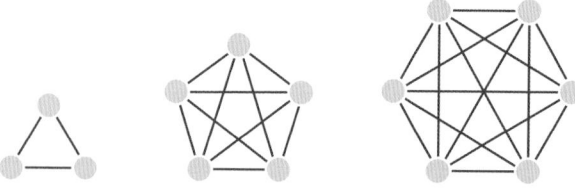

FIGURE 8.28 A clique graph consisting of three cliques.

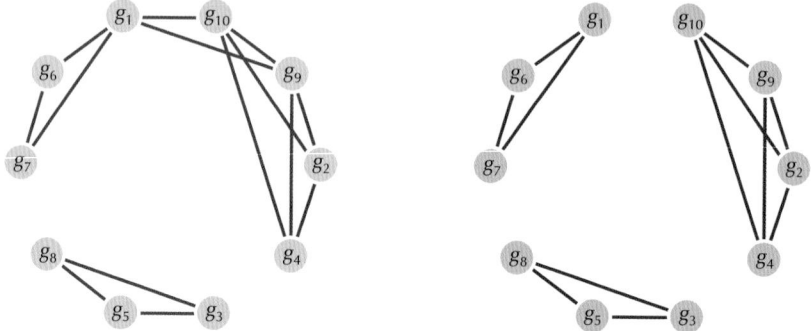

FIGURE 8.29 (Left) One possible similarity graph for the genes from Figure 8.22. (Right) The similarity graph can be transformed into a clique graph (right) by removing edges (g_1, g_{10}) and (g_1, g_9).

Corrupted Cliques Problem:

Find the minimum number of edges that need to be added or deleted to transform a graph into a clique graph.

   Input: A graph.
   Output: The minimum number of edge additions and deletions that transform this graph into a clique graph.

HOW DID YEAST BECOME A WINE MAKER?

The Corrupted Cliques Problem is difficult to solve exactly, so some heuristics have been proposed. The **Cluster Affinity Search Technique algorithm (CAST)**, described below, performs remarkably well at clustering gene expression data.

Define the similarity between gene i and cluster C as the average similarity between i and all genes in C:

$$R_{i,C} = \sum_{\text{all elements } j \text{ in cluster } C} \frac{R_{i,j}}{|C|}$$

Given a threshold θ, a gene i is θ-**close** to cluster C if $R_{i,C} > \theta$ and θ-**distant** from C otherwise. A cluster is called **consistent** if all genes in C are θ-close to C and all genes not in C are θ-distant from C. **CAST** uses the similarity graph and the threshold θ to iteratively find consistent clusters by starting with a single-element cluster C and then adding the "closest" gene not in C and removing the "most distant" gene in C. After a consistent cluster is found, all nodes in cluster C are removed from the similarity graph, and **CAST** iterates over the resulting smaller graph.

```
CAST(R, ϑ)
    Graph ← G(R, ϑ)
    Clusters ← empty set
    while Graph is nonempty
        C ← a single-node cluster consisting of a node of maximal degree in Graph
        while there exists a ϑ-close gene i not in C or a ϑ-distant gene i in C
            find the nearest ϑ-close gene i not in C and add it to C
            find the farthest ϑ-distant gene i in C and remove it from C
        add C to the set Clusters
        remove the nodes of C from Graph
    return Clusters
```

EXERCISE BREAK: Implement **CAST** and use it to cluster the abridged gene expression dataset.

CHAPTER 8

Bibliography Notes

The soft k-means algorithm for clustering, developed by Bezdek, 1981, is a variant of the expectation maximization algorithm, which was first proposed by Ceppellini, Siniscalco, and Smith, 1955 and rediscovered many times by various researchers. Do and Batzoglou, 2008 wrote an excellent primer on expectation maximization that inspired our discussion of coin flipping. The Lloyd algorithm for k-means clustering was introduced by Lloyd, 1982. Arthur and Vassilvitskii, 2007 developed k-means++ initialization for k-means clustering. The **CAST** algorithm was developed by Ben-Dor, Shamir, and Yakhini, 1999.

DeRisi, Iyer, and Brown, 1997 performed the first large-scale gene expression experiment to analyze the diauxic shift (see Cristianini and Hahn, 2007 for an excellent analysis of this experiment). Eisen et al., 1998 described the first applications of hierarchical clustering to gene expression analyses. Alon et al., 1999 analyzed patterns of gene expression in colon tumors.

Ohno, 1970 proposed the Whole Genome Duplication Model. Wolfe and Shields, 1997 provided the first convincing arguments in favor of a whole genome duplication in yeast. Kellis, Birren, and Lander, 2004 provided further evidence for a whole genome duplication by analyzing various yeast species. However, these arguments have not convinced Martin et al., 2007, who published a rebuttal. Thomson et al., 2005 resurrected the sequence of ancient alcohol dehydrogenases from yeast.

HOW DID YEAST BECOME A WINE MAKER?

How Do We Locate Disease-Causing Mutations?

What Causes Ohdo Syndrome?

About 1% of babies are born with mental retardation, but this affliction remains poorly understood because it can be caused by a variety of different genetic disorders. One of these disorders is **Ohdo syndrome**, which causes an expressionless, "mask-like" face. In 2011, biologists solved the genetic puzzle underlying Ohdo syndrome by discovering a handful of mutations shared by multiple patients, which the researchers used to identify a single protein-truncating mutation responsible for Ohdo syndrome.

The discovery of Ohdo syndrome's root cause represents just one of many new discoveries arising from the use of **read mapping** to study genetic disorders. In read mapping, researchers compare sequenced DNA reads taken from an individual against a **reference human genome** (see DETOUR: The Reference Human Genome) in order to find which reads perfectly match the reference and which reads indicate mutations of one nucleotide into another (**single nucleotide polymorphisms**, or **SNPs**). The reference genome is a gross simplification of species identity, since in addition to about 3 million SNPs (0.1% of the human genome), humans differ by genome rearrangements, insertions, and deletions that can span thousands of nucleotides (see DETOUR: Rearrangements, Insertions, and Deletions in Human Genomes). However, in this chapter we will focus only on algorithms for finding SNPs.

But wait, you may say, *why not use one of the algorithms that we already covered?* After all, we could always sequence the entire genome of an individual and then compare it against the reference genome. However, sequencing methods are computationally intensive and not perfect, as they often generate error-prone contigs. As a result, it makes sense to map reads from an individual human to the reference human genome to find out the differences.

To see why read mapping should be easier than genome assembly, let us return to the analogy of a jigsaw puzzle, which is sold with a picture of the completed puzzle on its box. This photo makes reconstructing the puzzle far easier; for a simple example, if the completed puzzle shows a sun in a blue sky, then you can automatically move all of the bright yellow pieces and all of the light blue pieces to the top of the puzzle.

Yet aside from genome sequencing, two other methods come to mind for read mapping. First, you could align each read to the reference genome (using a fitting alignment, Chapter 5) to find the most similar region. Second, you could apply approximate pattern matching algorithms to match each read one at a time against the reference genome.

STOP and Think: What computational challenges might arise from using these methods to map millions of reads to a reference human genome?

Both of these methods are guaranteed to solve the problem of mapping reads to a reference genome, but their runtimes become a bottleneck when we scale to millions of reads. Therefore, our goal in this chapter is to figure out how to use the reference genome as a "photo on the box" shortcut to find SNPs.

Introduction to Multiple Pattern Matching

Recall from Chapter 3 that reads are typically a few hundred base pairs long. These reads will form a collection of strings *Patterns* that we wish to match against a genome *Text*. For each string in *Patterns*, we will first find all its *exact* matches as a substring of *Text* (or conclude that it does not appear in *Text*). When hunting for the cause of a genetic disorder, we can immediately eliminate from consideration areas of the reference genome where exact matches occur. In the epilogue, we will generalize this problem to find *approximate* matches, where single nucleotide substitutions in reads separate the individual from the reference genome (or represent errors in reads).

Multiple Pattern Matching Problem:
Find all occurrences of a collection of patterns in a text.

> **Input**: A string *Text* and a collection *Patterns* containing (shorter) strings.
> **Output**: All starting positions in *Text* where a string from *Patterns* appears as a substring.

A naive approach to the Multiple Pattern Matching Problem would attempt repeated applications of an algorithm for the (single) Pattern Matching Problem, which we encountered in Chapter 1. This algorithm, which we call **BruteForcePatternMatching**, would slide each *Pattern* along *Text*, checking whether the substring starting at each position of *Text* matches *Pattern*. Recall that the runtime of a naive algorithm for a single pattern is $\mathcal{O}(|Text| \cdot |Pattern|)$. Thus, the runtime of **BruteForcePatternMatching** for the Multiple Pattern Matching Problem is $\mathcal{O}(|Text| \cdot |Patterns|)$, where $|Text|$ is the length of *Text* and $|Patterns|$ is the sum of the lengths of all strings in *Patterns*.

The problem with applying **BruteForcePatternMatching** to read mapping is that $|Text|$ and $|Patterns|$ are both huge. In the case of the human genome (3 GB), the total length of all reads may exceed 1 TB; as a result, any algorithm with runtime $\mathcal{O}(|Text| \cdot |Patterns|)$ will be too slow.

How Do We Locate Disease-Causing Mutations?

> **STOP and Think:** The estimate $\mathcal{O}(|Text| \cdot |Patterns|)$ presents the *worst-case* estimate of the runtime for **BruteForcePatternMatching**. What is the *average-case* estimate?

Herding Patterns into a Trie

Constructing a trie

The reason why the runtime of **BruteForcePatternMatching** is so high is that each string in *Patterns* must traverse all of *Text* independently. If you think about *Text* as a long road, then **BruteForcePatternMatching** is analogous to loading each pattern into its own car when driving down *Text*, an inefficient strategy. Instead, our goal is to herd the patterns onto a bus so that we only need to make one trip from the beginning to the end of *Text*. In more formal terms, we would like to organize *Patterns* into a data structure to prevent multiple passes down *Text* and to reduce the runtime. To this end, we will consolidate *Patterns* into a directed acyclic graph called a **trie** (pronounced "try"), which is written Trie(*Patterns*) and has the following properties (Figure 9.1).

- The trie has a single root node with indegree 0, denoted *root*.
- Each edge of Trie(*Patterns*) is labeled with a letter of the alphabet.
- Edges leading out of a given node have distinct labels.
- Every string in *Patterns* is spelled out by concatenating the letters along some path from the root downward.
- Every path from the root to a **leaf**, or node with outdegree 0, spells a string from *Patterns*.

Trie Construction Problem:
Construct a trie from a collection of patterns.

    Input: A collection of strings *Patterns*.
    Output: Trie(*Patterns*).

CHAPTER 9

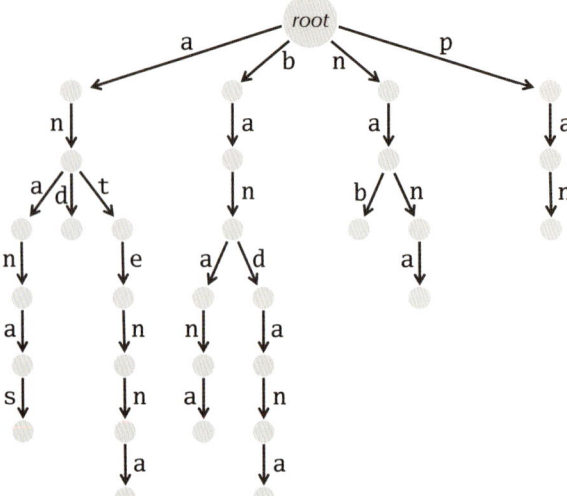

FIGURE 9.1 The trie for the following collection of strings *Patterns*: "ananas", "and", "antenna", "banana", "bandana", "nab", "nana", "pan".

The most obvious way to construct TRIE(*Patterns*) is by iteratively adding each string from *Patterns* to the growing trie, as implemented by the following algorithm.

TRIECONSTRUCTION(*Patterns*)
    Trie ← a graph consisting of a single node *root*
    for each string *Pattern* in *Patterns*
        currentNode ← *root*
        for $i ← 1$ to $|Pattern|$
            currentSymbol ← i-th symbol of *Pattern*
            if there is an outgoing edge from *currentNode* with label *currentSymbol*
                currentNode ← ending node of this edge
            else
                add a new node *newNode* to *Trie*
                add a new edge from *currentNode* to *newNode* with label *currentSymbol*
                currentNode ← *newNode*
    return *Trie*

STOP and Think: How can we use the trie to solve the Multiple Pattern Matching Problem?

Applying the trie to multiple pattern matching

Given a string *Text* and TRIE(*Patterns*), we can quickly check whether any string from *Patterns* matches a *prefix* of *Text*. To do so, we start reading symbols from the beginning of *Text* and see what string these symbols "spell" as we proceed along the path downward from the root of the trie, as illustrated in Figure 9.2 (left). For each new symbol in *Text*, if we encounter this symbol along an edge leading down from the present node, then we continue along this edge; otherwise, we stop and conclude that no string in *Patterns* matches a prefix of *Text*. If we make it all the way to a leaf, then the pattern spelled out by this path matches a prefix of *Text*. This algorithm is called **PREFIXTRIEMATCHING**.

PREFIXTRIEMATCHING(*Text*, *Trie*)
    symbol ← first letter of *Text*
    v ← root of *Trie*
    while forever
        if *v* is a leaf in *Trie*
            return the pattern spelled by the path from the root to *v*
        else if there is an edge (*v*, *w*) in *Trie* labeled by *symbol*
            symbol ← next letter of *Text*
            v ← *w*
    else
        output "no matches found"
        return

STOP and Think: For **PREFIXTRIEMATCHING** to work, we have made a hidden assumption that no string in *Patterns* is a prefix of another string in *Patterns*. How can this algorithm be modified when *Patterns* is an arbitrary collection of strings? Hint: consider adding "pantry" to the patterns in Figure 9.2 (left).

PREFIXTRIEMATCHING finds whether any strings in *Patterns* match a prefix of *Text*. To find whether any strings in *Patterns* match a substring of *Text* starting at position *k*, we chop off the first *k* − 1 symbols from *Text* and run **PREFIXTRIEMATCHING** on the shortened string. As a result, to solve the Multiple Pattern Matching Problem, we simply iterate **PREFIXTRIEMATCHING** |*Text*| times, chopping the first symbol off of *Text* before each new iteration (Figure 9.2 (right)).

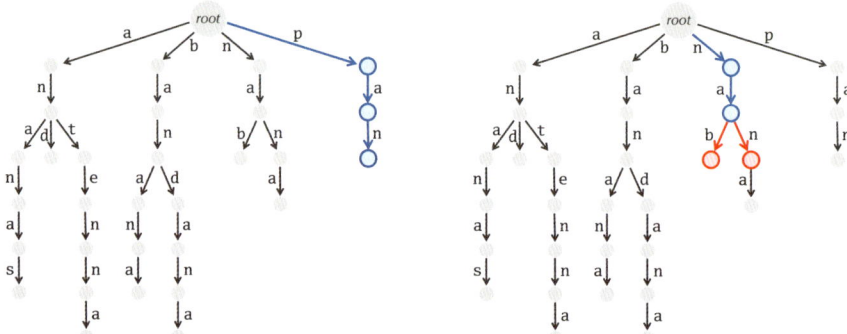

FIGURE 9.2 (Left) The pattern "pan" matches Text = "**pan**amabananas" when starting at the beginning of *Text*. (Right) No pattern match is found in TRIE(*Patterns*) for the strings *Patterns* from Figure 9.1 when starting at the third symbol of "pa**na**mabananas".

    TRIEMATCHING(Text, Trie)
        while Text is nonempty
            PREFIXTRIEMATCHING(Text, Trie)
            remove first symbol from Text

We need |*Patterns*| steps to construct TRIE(*Patterns*), which contains at most |*Patterns*| nodes. Each iteration of **PREFIXTRIEMATCHING** takes at most |*LongestPattern*| steps, where *LongestPattern* is the longest string in *Patterns*. **TRIEMATCHING** makes |*Text*| total calls to **PREFIXTRIEMATCHING**, making the total number of steps equal to |*Patterns*| + |*Text*| · |*LongestPattern*|. This runtime offers a huge speed-up compared to the |*Text*| · |*Patterns*| steps required by **BRUTEFORCEPATTERNMATCHING**. The **Aho-Corasick algorithm**, developed in 1975, further reduces the number of steps required after constructing the trie from $\mathcal{O}(|Text| \cdot |LongestPattern|)$ steps to $\mathcal{O}(|Text|)$ steps (see **DETOUR: The Aho-Corasick Algorithm**).

STOP and Think: Do you see any computational challenges with using **TRIEMATCHING** to solve the Multiple Pattern Matching Problem?

Although **TRIEMATCHING** is fast, storing a trie consumes a lot of memory. Recall that **BRUTEFORCEPATTERNMATCHING** works with a single read at a time, which keeps the memory low because we only need to store the genome in memory. Yet **TRIEMATCHING** needs to store the entire trie in memory, which is proportional to

HOW DO WE LOCATE DISEASE-CAUSING MUTATIONS?

|*Patterns*|. Since a collection of reads for the human genome may consume upwards of 1 TB, the memory required to store the trie is prohibitive.

STOP and Think: How can we avoid multiple passes through the genome without needing to consolidate all the reads into a huge data structure?

Preprocessing the Genome Instead

Introduction to suffix tries

Since storing TRIE(*Patterns*) requires so much memory, let's process *Text* into a data structure instead. Our goal is to compare each string in *Patterns* against *Text* without needing to traverse *Text* from beginning to end. In more familiar terms, instead of packing *Patterns* onto a bus and riding the long distance down *Text*, our new data structure will be able to "teleport" each string in *Patterns* directly to its occurrences in *Text*.

A **suffix trie**, denoted SUFFIXTRIE(*Text*), is the trie formed from all suffixes of *Text* (Figure 9.3). From now on, we append the dollar-sign ("$") to *Text* in order to mark the end of *Text*. We will also label each leaf of the resulting trie by the starting position of the suffix whose path through the trie ends at this leaf (using 0-based indexing). This way, when we arrive at a leaf, we will immediately know where this suffix came from in *Text*.

STOP and Think: How can we use the suffix trie for pattern matching?

Using suffix tries for pattern matching

To match a single string *Pattern* to *Text*, note that if *Pattern* matches a substring of *Text* starting at position *i*, then *Pattern* must also appear at the beginning of the suffix of *Text* starting at position *i*. We can therefore determine whether *Pattern* occurs in SUFFIXTRIE(*Text*) by starting at the root and spelling symbols of *Pattern* downward. If we can find a path in the suffix trie spelling out *Pattern*, then we know that *Pattern* must occur in *Text* (Figure 9.4). We can then iterate over all strings in *Patterns*.

STOP and Think: Figure 9.4 illustrates how to find the pattern "nanas" in SUFFIXTRIE("panamabananas$"), but it does not tell us where "nanas" occurs in *Text*. How can we obtain this information?

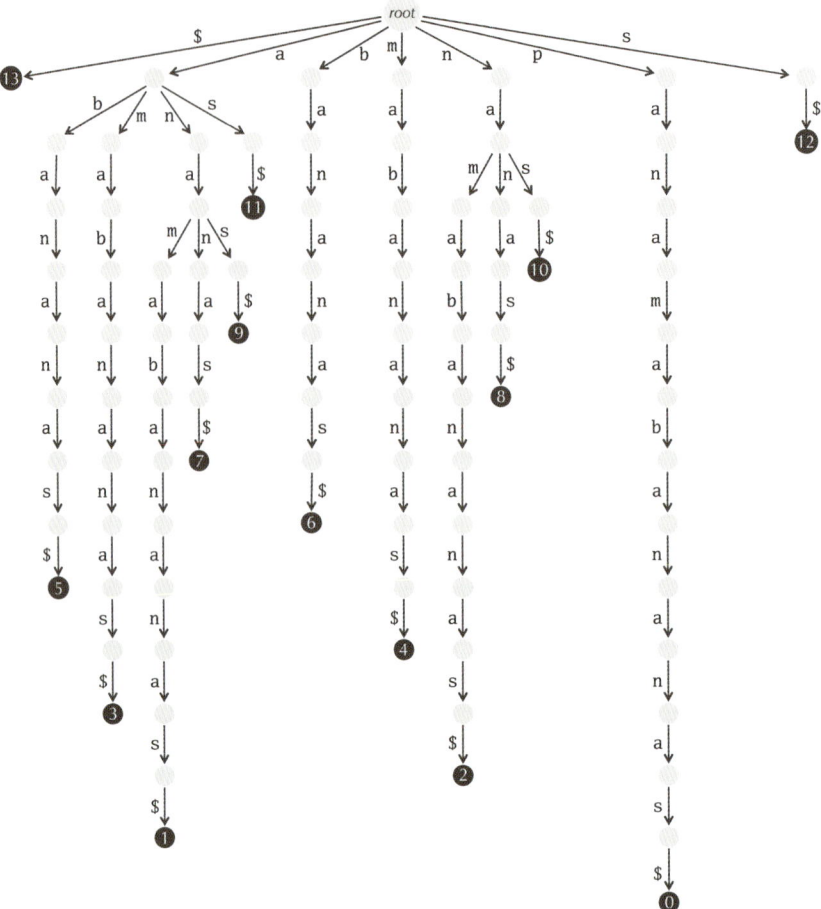

FIGURE 9.3 SUFFIXTRIE("panamabananas$"), with leaf labels (corresponding to starting positions of suffixes) varying from 0 to 13.

To determine where *Pattern* appears in *Text*, assume first that *Pattern* matches *Text* at a leaf of SUFFIXTRIE(*Text*). In this case, *Pattern* must appear in *Text* as a suffix, and we can consult the label at that leaf to determine the starting position of the suffix. For example, threading "nanas" into the suffix trie in Figure 9.4 shows that it matches the suffix starting at position 8 of "panamaba**nanas**$".

If the path spelling out *Pattern* stops before a leaf at some node v of SUFFIXTRIE(*Text*), then *Pattern* may occur more than once in *Text*. To locate these occurrences, follow all paths from v down to the leaves of SUFFIXTRIE(*Text*), which will indicate all starting

HOW DO WE LOCATE DISEASE-CAUSING MUTATIONS?

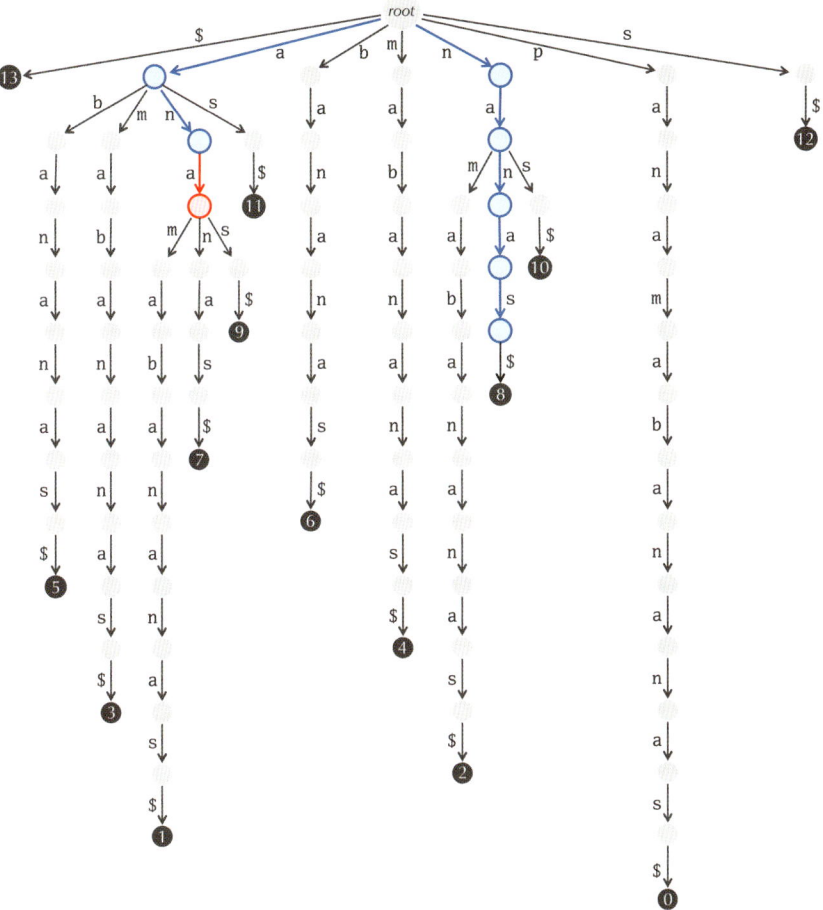

FIGURE 9.4 Threading "antenna" through SUFFIXTRIE("panamabananas$") fails to find a match because no suffix of "panamabananas$" begins with "**ant**"; however, threading "nana" through the suffix trie does find a match: "panamaba**nana**s$".

positions of *Pattern* in *Text*. For example, the pattern "ana" corresponds to a path in SUFFIXTRIE("panamabananas$") that can be extended to three different leaves with labels 1, 7, and 9, corresponding to three occurrences of "ana": "p**ana**mabananas$", "panamab**ana**nas$", and "panamaban**ana**s$".

STOP and Think: How much runtime and memory will it take to construct SUFFIXTRIE(*Text*)?

CHAPTER 9

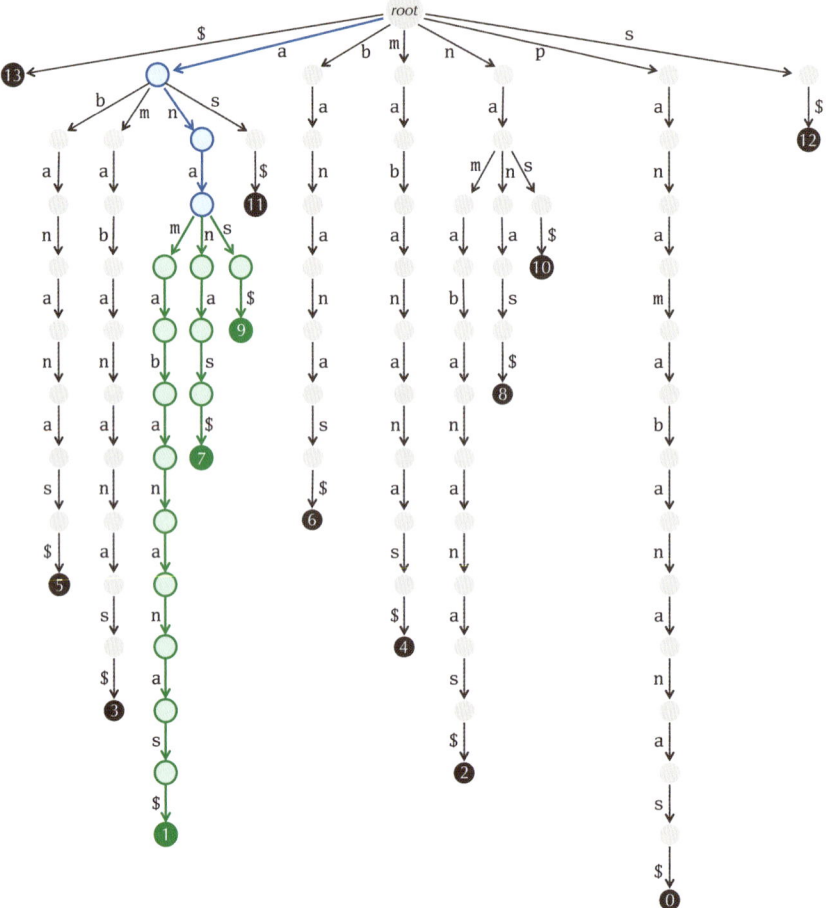

FIGURE 9.5 All paths starting with "ana" reveal the three occurrences of "ana" in "panamabananas$". Extending these paths to the leaves (shown in green) reveals that the starting positions of these occurrences are 1, 7, and 9.

Recall that constructing TRIE(*Patterns*) required $\mathcal{O}(|Patterns|)$ runtime and memory. Accordingly, the runtime and memory required to construct SUFFIXTRIE(*Text*) are both equal to the combined length of all suffixes in *Text*. There are $|Text|$ suffixes of *Text*, ranging in length from 1 to $|Text|$ and having total length $|Text| \cdot (|Text| + 1)/2$, which is $\mathcal{O}(|Text|^2)$. Thus, we need to reduce both the construction time and memory requirements of suffix tries to make them practical.

Suffix Trees

Let's not give up hope on suffix tries. We can reduce the number of edges in the suffix trie by combining the edges on any non-branching path into a single edge. We then label this edge with the concatenation of symbols on the consolidated edges, as shown in Figure 9.6. The resulting data structure is called a **suffix tree**, written SUFFIXTREE(*Text*).

To match a single *Pattern* to *Text*, we thread *Pattern* into SUFFIXTREE(*Text*) by the same process used for a suffix trie. Similarly to the suffix trie, we can use the leaf labels to find starting positions of successfully matched patterns.

EXERCISE BREAK: Prove that SUFFIXTREE(*Text*) has exactly |*Text*| + 1 leaves and at most |*Text*| + 1 other nodes.

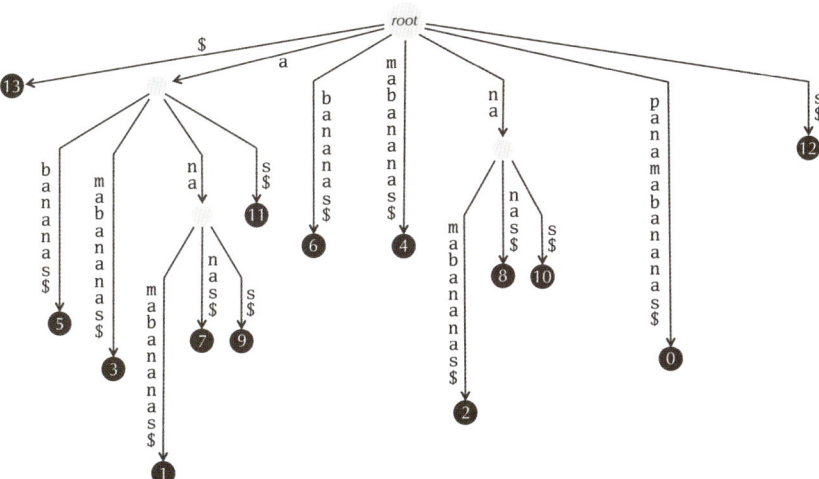

FIGURE 9.6 SUFFIXTREE("panamabananas$"), formed by compressing the edges on non-branching paths in Figure 9.3.

Compared to a suffix trie, which may have a quadratic number of nodes in the length of *Text*, the number of nodes in SUFFIXTREE(*Text*) is at most 2 · |*Text*|. Therefore, the memory required for SUFFIXTREE(*Text*) is $\mathcal{O}(|Text|)$.

STOP and Think: Wait a second. Suffix trees seem like a cosmetic modification of suffix tries. Since we still need to keep all concatenated edge labels in memory, why should suffix trees be more memory-efficient than suffix tries?

CHAPTER 9

Suffix trees save memory because they do not need to store concatenated edge labels from each non-branching path. For example, a suffix tree does not need ten bytes to store the edge labeled "mabananas$" in Figure 9.6; instead, it suffices to store a **pointer** to position 4 of "panamabananas$", as well as the *length* of "mabananas$". Furthermore, suffix trees can be constructed in linear time, without having to first construct the suffix trie! We will not ask you to implement this fast suffix tree construction algorithm because it is quite complex.

Suffix Tree Construction Problem:
Construct the suffix tree of a string.

> **Input**: A string *Text*.
> **Output**: SUFFIXTREE(*Text*).

CHARGING STATION (Constructing a Suffix Tree): A memory-inefficient way of constructing the suffix tree is to first construct the suffix trie and then consolidate each non-branching path into a single edge, storing its label in memory. To implement a more memory-efficient solution, check out this Charging Station.

Although the suffix tree decreases memory requirements from $\mathcal{O}(|Text|^2)$ to $\mathcal{O}(|Text|)$, on average it still requires about 20 times as much memory as *Text*. In the case of a 3 GB human genome, 60 GB of RAM is a huge improvement over the 1 TB that we needed to work with TRIE(*Patterns*), but it still presents a memory challenge for most machines. This reveals a dark secret of big-O notation, which is that it ignores constant factors. For long strings such as the human genome, we will need to pay attention to this constant factor, since the expression $\mathcal{O}(|Text|)$ applies to both an algorithm with $2 \cdot |Text|$ memory and an algorithm with $1000 \cdot |Text|$ memory.

Yet before seeing how we can further reduce the memory needed for multiple pattern matching, we ask you to solve three problems for which suffix trees are useful.

Longest Repeat Problem:
Find the longest repeat in a string.

> **Input**: A string *Text*.
> **Output**: A longest substring of *Text* that appears in *Text* more than once.

HOW DO WE LOCATE DISEASE-CAUSING MUTATIONS?

Longest Shared Substring Problem:

Find the longest substring shared by two strings.

    Input: Strings $Text_1$ and $Text_2$.
    Output: The longest substring that occurs in both $Text_1$ and $Text_2$.

Shortest Non-Shared Substring Problem:

Find the shortest substring of one string that does not appear in another string.

    Input: Strings $Text_1$ and $Text_2$.
    Output: The shortest substring of $Text_1$ that does not appear in $Text_2$.

CHARGING STATION (Solving the Longest Shared Substring Problem):
One way of solving the Longest Shared Substring Problem is to construct two suffix trees, one for $Text_1$ and one for $Text_2$. Check out this Charging Station to learn about a more elegant solution.

Suffix Arrays

Constructing a suffix array

In 1993, Udi Manber and Gene Myers introduced **suffix arrays** as a memory-efficient alternative to suffix trees. To construct SUFFIXARRAY(*Text*), we first sort all suffixes of *Text* lexicographically, assuming that "$" comes first in the alphabet (Figure 9.7). The suffix array is the list of starting positions of these sorted suffixes:

    SUFFIXARRAY("panamabananas$") = [13, 5, 3, 1, 7, 9, 11, 6, 4, 2, 8, 10, 0, 12].

Suffix Array Construction Problem:

Construct the suffix array of a string.

    Input: A string *Text*.
    Output: SUFFIXARRAY(*Text*).

CHAPTER 9

Sorted Suffixes	Starting Positions
$	13
abananas$	5
amabananas$	3
anamabananas$	1
ananas$	7
anas$	9
as$	11
bananas$	6
mabananas$	4
namabananas$	2
nanas$	8
nas$	10
panamabananas$	0
s$	12

FIGURE 9.7 The sorted list of suffixes of *Text* = "panamabananas$", along with their starting positions in *Text*; these starting positions form the suffix array of *Text*.

The Suffix Array Construction Problem can easily be solved after sorting all suffixes of *Text*, but since even the fastest algorithms for sorting an array of n elements require $\mathcal{O}(n \cdot \log n)$ comparisons, sorting all suffixes takes $\mathcal{O}(|Text| \cdot \log(|Text|))$ comparisons. However, there exists a faster algorithm that constructs suffix arrays in linear time and requires only about a fifth as much memory as suffix trees, which knocks the 60 GB memory requirement for the human genome down to 12 GB.

STOP and Think: Given a suffix tree, can you quickly transform it into a suffix array? Given a suffix array, can you quickly transform it into a suffix tree?

As the preceding question suggests, suffix arrays and suffix trees are practically equivalent; every algorithm using suffix trees can be translated into an algorithm using suffix arrays (see **DETOUR: From Suffix Trees to Suffix Arrays**), and vice-versa (see **DETOUR: From Suffix Arrays to Suffix Trees**).

Pattern matching with the suffix array

Once we have constructed the suffix array of a string *Text*, we can use it to quickly locate every occurrence of a string *Pattern* in *Text*. First, recall that when pattern matching with the suffix trie, we observed that all matches of *Pattern* in *Text* must occur at the beginning of suffixes of *Text*. Second, note that after sorting the suffixes of *Text*, the

How do we locate disease-causing mutations?

suffixes beginning with *Pattern* clump together. For example, in Figure 9.7, *Pattern* = "ana" occurs at the beginning of the suffixes "anamabananas$", "ananas$", and "anas$" of *Text* = "panamabananas$"; these suffixes occur in three consecutive rows and correspond to the starting positions 1, 7, and 9 in *Text*.

The question is how to find these starting positions for an arbitrary string *Pattern* without needing to store the sorted suffixes of *Text*. The following algorithm, called **PatternMatchingWithSuffixArray**, identifies the first and last index of the suffix array corresponding to suffixes beginning with *Pattern* (these indices are denoted *first* and *last*, respectively). **PatternMatchingWithSuffixArray** offers a variation of a general search technique called **binary search** that finds a data point in a sorted collection of data by iteratively dividing the data in half and determining the half in which the data point lies. See **DETOUR: Binary Search** for more details.

PatternMatchingWithSuffixArray(*Text*, *Pattern*, SuffixArray)
    minIndex ← 0
    maxIndex ← |*Text*|
    while *minIndex* < *maxIndex*
        midIndex ← (*minIndex* + *maxIndex*)/2
        if *Pattern* > suffix of *Text* starting at position SuffixArray(*midIndex*)
            minIndex ← *midIndex* + 1
        else
            maxIndex ← *midIndex*
    first ← *minIndex*
    maxIndex ← |*Text*|
    while *minIndex* < *maxIndex*
        midIndex ← (*minIndex* + *maxIndex*)/2
        if *Pattern* < suffix of *Text* starting at position SuffixArray(*midIndex*)
            maxIndex ← *midIndex*
        else
            minIndex ← *midIndex* + 1
    last ← *maxIndex*
    if *first* > *last*
        return "*Pattern* does not appear in *Text*"
    else
        return (*first*, *last*)

CHAPTER 9

The Burrows-Wheeler Transform

Genome compression

Suffix arrays have greatly reduced the memory required for efficient text searches, and until the start of this century, they represented the state of the art in pattern matching. Can we be so ambitious as to look for a data structure that would encode *Text* using memory approximately equal to the length of *Text* while still enabling fast pattern matching?

To answer this question, we will digress to consider the seemingly unrelated topic of **text compression**. In one simple compression technique called **run-length encoding**, we replace a **run** of *k* consecutive occurrences of symbol *s* with only two symbols: *k*, followed by *s*. For example, run-length encoding would compress the string `TTTTTGGGAAAACCCCCCA` into `5T3G4A6C1A`.

Run-length encoding works well for strings having lots of long runs, but real genomes do not have many runs. What they do have, as we saw in Chapter 3, are *repeats*. It would therefore be nice if we could first manipulate the genome to convert repeats into runs and then apply run-length encoding to the resulting string.

A naive way of creating runs in a string is to reorder the string's symbols lexicographically. For example, `TACGTAACGATACGAT` would become `AAAAACCCGGGTTTT`, which we could then compress into `5A3C3G4T`. This method would represent a 3 GB human genome file using just four numbers.

STOP and Think: What is wrong with applying this compression method to genomes?

Ordering a string's symbols lexicographically is not suitable for compression because many different strings will get compressed into the *same* string. For example, the DNA strings `GCATCATGCAT` and `ACTGACTACTG` — as well as any string with the same nucleotide counts — get reordered into `AAACCCGGTTT`. As a result, we cannot **decompress** the compressed string, i.e., invert the compression operation to produce the original string.

Constructing the Burrows-Wheeler transform

Let's consider a different method of converting the repeats of a string into runs that was proposed by Michael Burrows and David Wheeler in 1994. First, form all possible **cyclic rotations** of *Text*; a cyclic rotation is defined by chopping off a suffix from the end of

Text and appending this suffix to the beginning of *Text*. Next — similarly to suffix arrays — order all the cyclic rotations of *Text* lexicographically to form a |*Text*| × |*Text*| matrix of symbols that we call the **Burrows-Wheeler matrix** and denote by M(*Text*) (Figure 9.8).

Cyclic Rotations	M("panamabananas$")
panamabananas$	$ p a n a m a b a n a n a **s**
$panamabananas	a b a n a n a s $ p a n a **m**
s$panamabanana	a m a b a n a n a s $ p a **n**
as$panamabanan	a n a m a b a n a n a s $ **p**
nas$panamabana	a n a n a s $ p a n a m a **b**
anas$panamaban	a n a s $ p a n a m a b a **n**
nanas$panamaba	a s $ p a n a m a b a n a **n**
ananas$panamab	b a n a n a s $ p a n a m **a**
bananas$panama	m a b a n a n a s $ p a n **a**
abananas$panam	n a m a b a n a n a s $ p **a**
mabananas$pana	n a n a s $ p a n a m a b **a**
amabananas$pan	n a s $ p a n a m a b a n **a**
namabananas$pa	p a n a m a b a n a n a s **$**
anamabananas$p	s $ p a n a m a b a n a n **a**

FIGURE 9.8 All cyclic rotations of "panamabananas$" (left) and the Burrows-Wheeler matrix M("panamabananas$") of all lexicographically ordered cyclic rotations (right). BWT("panamabananas$") is the last column of M("panamabananas$"): "**smnpbnnaaaaa$a**".

Notice that the first column of M(*Text*) contains the symbols of *Text* ordered lexicographically, which is just the naive rearrangement of *Text* that we already described. In turn, the second column of M(*Text*) contains the second symbols of all cyclic rotations of *Text*, and so it too represents a (different) rearrangement of symbols from *Text*. The same reasoning applies to show that any column of M(*Text*) is some rearrangement of the symbols of *Text*. We are interested in the last column of M(*Text*), called the **Burrows-Wheeler transform** of *Text*, or BWT(*Text*), which is shown in red in Figure 9.8.

STOP and Think: We have seen that the first column of M(*Text*) cannot be uniquely decompressed to yield *Text*. Do you think that some other column of M(*Text*) can be inverted to yield *Text*?

CHAPTER 9

Burrows-Wheeler Transform Construction Problem:

Construct the Burrows-Wheeler transform of a string.

Input: A string *Text*.
Output: BWT(*Text*).

STOP and Think: Figure 9.8 suggests a simple algorithm for computing BWT(*Text*) based on constructing M(*Text*). Can you construct BWT(*Text*) using less memory given *Text* and SUFFIXARRAY(*Text*)?

From repeats to runs

If we re-examine the Burrows-Wheeler transform in Figure 9.8, we immediately notice that it has created the run "aaaaa" in BWT("panamabananas$") = "smnpbnn**aaaaa**$a".

STOP and Think: Why do you think that the Burrows-Wheeler Transform produced this run?

Imagine that we take the Burrows-Wheeler transform of Watson and Crick's 1953 paper on the double helix structure of DNA. The word "and" is repeated often in English, which means that when we form all possible cyclic rotations of the Watson & Crick paper, we will witness a large number of rotations beginning with "and...". In turn, we will observe many rotations that begin with "nd..." and end with "...a". When all the cyclic rotations of *Text* are sorted lexicographically to form M(*Text*), all rows that begin with "nd..." and end with "...a" will tend to clump together. As illustrated in Figure 9.9, this clumping produces runs of "a" in the final column of M(*Text*), which we know is BWT(*Text*).

The substring "ana" in "panamabananas$" plays the role of "and" in Watson and Crick's paper and explains three of the five occurrences of "a" in the repeat "aaaaa" in BWT("panamabananas$") = "smnpbnn**aaaaa**$a". When the Burrows-Wheeler transform is applied to a genome, it converts the genome's many repeats into runs. As we already suggested, after applying the Burrows-Wheeler transform, we can apply an additional compression method such as run-length encoding in order to further reduce the memory.

HOW DO WE LOCATE DISEASE-CAUSING MUTATIONS?

```
nd Corey (1). They kindly made their manuscript availa ...... a
nd criticism, especially on interatomic distances. We    ...... a
nd cytosine. The sequence of bases on a single chain d   ...... a
nd experimentally (3,4) that the ratio of the amounts o  ...... u
nd for this reason we shall not comment on it. We wish   ...... a
nd guanine (purine) with cytosine (pyrimidine). In oth   ...... a
nd ideas of Dr. M. H. F. Wilkins, Dr. R. E. Franklin     ...... a
nd its water content is rather high. At lower water co   ...... a
nd pyrimidine bases. The planes of the bases are perpe   ...... a
nd stereochemical arguments. It has not escaped our no   ...... a
nd that only specific pairs of bases can bond together   ...... u
nd the atoms near it is close to Furberg's 'standard co  ...... a
nd the bases on the inside, linked together by hydrogen  ...... a
nd the bases on the outside. In our opinion, this stru   ...... a
nd the other a pyrimidine for bonding to occur. The hy   ...... a
nd the phosphates on the outside. The configuration of   ...... a
nd the ration of guanine to cytosine, are always very c  ...... a
nd the same axis (see diagram). We have made the usual   ...... u
nd their co-workers at King's College, London. One of    ...... a
```

FIGURE 9.9 A few consecutive rows selected from M(*Text*), where *Text* is Watson and Crick's 1953 paper on the double helix. Rows beginning with "**nd...**" often end with "**...a**" because of the common occurrence of the word "and" in English, which causes runs of "**a**" in BWT(*Text*).

EXERCISE BREAK: There is only one run of length at least 10 in the *E. coli* genome. How many runs of length at least 10 do you find after applying the Burrows-Wheeler transform to the *E. coli* genome?

Inverting the Burrows-Wheeler Transform

A first attempt at inverting the Burrows-Wheeler transform

Before we get ahead of ourselves, remember that compressing a genome does not count for much if we cannot decompress it. In particular, if there exist a pair of genomes that the Burrows-Wheeler transform compresses into the same string, then we will not be able to decompress this string. But it turns out that the Burrows-Wheeler transform is reversible!

STOP and Think: Can you find the (unique) string whose Burrows-Wheeler transform is "enwvpeoseu$llt"? It could be "newtloveslupe$", "elevenplustwo$", "unwellpesovet$", or something else entirely.

CHAPTER 9

Consider the toy example BWT(*Text*) = "ard$rcaaaabb". First, recall that the first column of M(*Text*) is the lexicographic rearrangement of symbols in BWT(*Text*), i.e., "$aaaaabbcdrr". For convenience, we will use the terms *FirstColumn* and *LastColumn* (i.e., BWT(*Text*)) when referring to the first and last columns of M(*Text*), respectively.

We know that the first row of M(*Text*) is the cyclic rotation of *Text* beginning with "$", which occurs at the end of *Text*. Thus, if we determine the first row of M(*Text*), then we can move the "$" to the end of this row and reproduce *Text*. But how do we determine the remaining symbols in this first row, if all we know is *FirstColumn* and *LastColumn*?

```
$ ? ? ? ? ? ? ? ? ? a
a ? ? ? ? ? ? ? ? ? r
a ? ? ? ? ? ? ? ? ? d
a ? ? ? ? ? ? ? ? ? $
a ? ? ? ? ? ? ? ? ? r
a ? ? ? ? ? ? ? ? ? c
b ? ? ? ? ? ? ? ? ? a
b ? ? ? ? ? ? ? ? ? a
c ? ? ? ? ? ? ? ? ? a
d ? ? ? ? ? ? ? ? ? a
r ? ? ? ? ? ? ? ? ? b
r ? ? ? ? ? ? ? ? ? b
```

STOP and Think: Using the first and last columns of the Burrows-Wheeler matrix shown above, can you find the first symbol of *Text*?

Note that the first symbol in *Text* must follow "$" in *any* cyclic rotation of *Text*. Because "$" occurs as the fourth symbol of *LastColumn* = "ard$rcaaaabb", we know that if we walk one symbol to the right from the end of the fourth row of M(*Text*), then we will "wrap around" and arrive at the fourth symbol of *FirstColumn*, which is **"a"** in "$aa**a**aabbcdrr". Therefore, this **"a"** belongs in the first position of *Text*:

```
$ a ? ? ? ? ? ? ? ? a
a ? ? ? ? ? ? ? ? ? r
a ? ? ? ? ? ? ? ? ? d
a ? ? ? ? ? ? ? ? ? $
a ? ? ? ? ? ? ? ? ? r
a ? ? ? ? ? ? ? ? ? c
b ? ? ? ? ? ? ? ? ? a
b ? ? ? ? ? ? ? ? ? a
c ? ? ? ? ? ? ? ? ? a
d ? ? ? ? ? ? ? ? ? a
r ? ? ? ? ? ? ? ? ? b
r ? ? ? ? ? ? ? ? ? b
```

STOP and Think: Which symbol is hiding in the second position of *Text*?

Following the same logic of "wrapping around", the next symbol of *Text* should be the first symbol in a row of M(*Text*) that ends in "a". The only trouble is that five rows end in "a", and we don't know which of them is the correct one! If we guess that this "a" is the seventh symbol of "ard$rcaaaabb", then we obtain "b" in the second position of *Text* (Figure 9.10 (left)). On the other hand, if we guess that this "a" is the ninth symbol of "ard$rcaaaabb", then we obtain "c" in the second position of *Text* (Figure 9.10 (middle)). Finally, if we guess that this "a" is the tenth symbol of "ard$rcaaaabb", then we obtain "d" in the second position of *Text* (Figure 9.10 (right)).

```
$ a b ? ? ? ? ? ? ? ? a        $ a c ? ? ? ? ? ? ? ? a        $ a d ? ? ? ? ? ? ? ? a
a ? ? ? ? ? ? ? ? ? ? r        a ? ? ? ? ? ? ? ? ? ? r        a ? ? ? ? ? ? ? ? ? ? r
a ? ? ? ? ? ? ? ? ? ? d        a ? ? ? ? ? ? ? ? ? ? d        a ? ? ? ? ? ? ? ? ? ? d
a ? ? ? ? ? ? ? ? ? ? $        a ? ? ? ? ? ? ? ? ? ? $        a ? ? ? ? ? ? ? ? ? ? $
a ? ? ? ? ? ? ? ? ? ? r        a ? ? ? ? ? ? ? ? ? ? r        a ? ? ? ? ? ? ? ? ? ? r
a ? ? ? ? ? ? ? ? ? ? c        a ? ? ? ? ? ? ? ? ? ? c        a ? ? ? ? ? ? ? ? ? ? c
b ? ? ? ? ? ? ? ? ? ? a        b ? ? ? ? ? ? ? ? ? ? a        b ? ? ? ? ? ? ? ? ? ? a
b ? ? ? ? ? ? ? ? ? ? a        b ? ? ? ? ? ? ? ? ? ? a        b ? ? ? ? ? ? ? ? ? ? a
c ? ? ? ? ? ? ? ? ? ? a        c ? ? ? ? ? ? ? ? ? ? a        c ? ? ? ? ? ? ? ? ? ? a
d ? ? ? ? ? ? ? ? ? ? a        d ? ? ? ? ? ? ? ? ? ? a        d ? ? ? ? ? ? ? ? ? ? a
r ? ? ? ? ? ? ? ? ? ? b        r ? ? ? ? ? ? ? ? ? ? b        r ? ? ? ? ? ? ? ? ? ? b
r ? ? ? ? ? ? ? ? ? ? b        r ? ? ? ? ? ? ? ? ? ? b        r ? ? ? ? ? ? ? ? ? ? b
```

FIGURE 9.10 The three possibilities ("b", "c", or "d") for the third element of the first row of M(*Text*) when BWT(*Text*) is "ard$rcaaaabb". One of these possibilities must correspond to the second symbol of *Text*.

STOP and Think: How would you choose among "b", "c", and "d" for the second symbol of *Text*?

The First-Last Property

To determine the remaining symbols of *Text*, we need to use a subtle property of M(*Text*) that may seem completely unrelated to inverting the Burrows-Wheeler transform. Below, we have indexed the occurrences of each symbol in *FirstColumn* with subscripts according to their order of appearance in this column. When *Text* = "panamabananas$", six instances of "a" appear in *FirstColumn*.

CHAPTER 9

```
$  p a n a m a b a n a n a s
a₁ b a n a n a s $ p a n a m
a₂ m a b a n a n a s $ p a n
a₃ n a m a b a n a n a s $ p
a₄ n a n a s $ p a n a m a b
a₅ n a s $ p a n a m a b a n
a₆ s $ p a n a m a b a n a n
b  a n a n a s $ p a n a m a
m  a b a n a n a s $ p a n a
n  a m a b a n a n a s $ p a
n  a n a s $ p a n a m a b a
n  a s $ p a n a m a b a n a
p  a n a m a b a n a n a s $
s  $ p a n a m a b a n a n a
```

Consider "a_1" in *FirstColumn*, which occurs at the beginning of the cyclic rotation "$a_1$bananas$panam". If we cyclically rotate this string, then we obtain "panam$a_1$bananas$". Thus, "$a_1$" in *FirstColumn* is actually the third occurrence of "a" in "panamabananas$". We can now identify the positions of the other five instances of "a" in "panamabananas$":

$$pa_3na_2ma_1ba_4na_5na_6s\$$$

EXERCISE BREAK: Where are the three instances of "n" from *FirstColumn* (i.e., "n_1", "n_2", and "n_3") located in "panamabananas$"?

To locate "a_1" in *LastColumn*, we need to cyclically rotate the second row of the matrix M("$a_1$bananas$panam"), which results in "bananas$panam$a_1$". This rotation corresponds to the eighth row of M("panamabananas$"):

```
$  p a n a m a b a n a n a s
a₁ b a n a n a s $ p a n a m
a₂ m a b a n a n a s $ p a n
a₃ n a m a b a n a n a s $ p
a₄ n a n a s $ p a n a m a b
a₅ n a s $ p a n a m a b a n
a₆ s $ p a n a m a b a n a n
b  a n a n a s $ p a n a m a₁
m  a b a n a n a s $ p a n a
n  a m a b a n a n a s $ p a
n  a n a s $ p a n a m a b a
n  a s $ p a n a m a b a n a
p  a n a m a b a n a n a s $
s  $ p a n a m a b a n a n a
```

142

EXERCISE BREAK: Where are the other five instances of "a" located in *LastColumn*?

You hopefully saw that *LastColumn* can be recorded as "smnpbnna$_1$a$_2$a$_3$a$_4$a$_5$$a$_6$", as shown in Figure 9.11. Note that the six instances of "a" appear in exactly the same order in *FirstColumn* and *LastColumn*. This observation is not a fluke. On the contrary, it is a principle that holds for any string *Text* and any symbol that we choose.

First-Last Property: *The k-th occurrence of a symbol in FirstColumn and the k-th occurrence of this symbol in LastColumn correspond to the same position of this symbol in Text.*

```
$     p a n a m a b a n a n a s
a₁    b a n a n a s $ p a n a m
a₂    m a b a n a n a s $ p a n
a₃    n a m a b a n a n a s $ p
a₄    n a n a s $ p a n a m a b
a₅    n a s $ p a n a m a b a n
a₆    s $ p a n a m a b a n a n
b     a n a n a s $ p a n a m a₁
m     a b a n a n a s $ p a n a₂
n     a m a b a n a n a s $ p a₃
n     a n a s $ p a n a m a b a₄
n     a s $ p a n a m a b a n a₅
p     a n a m a b a n a n a s $
s     $ p a n a m a b a n a n a₆
```

FIGURE 9.11 The six occurrences of "a" occur in the same order in *FirstColumn* as they do in *LastColumn*.

To see why the First-Last Property is true, consider the rows of M("panamabananas$") beginning with "a":

```
a₁ b a n a n a s $ p a n a m
a₂ m a b a n a n a s $ p a n
a₃ n a m a b a n a n a s $ p
a₄ n a n a s $ p a n a m a b
a₅ n a s $ p a n a m a b a n
a₆ s $ p a n a m a b a n a n
```

These rows are already ordered lexicographically, so if we chop off the "a" from the beginning of each row, then the remaining strings should still be ordered lexicographically:

CHAPTER 9

```
b a n a n a s $ p a n a m
m a b a n a n a s $ p a n
n a m a b a n a n a s $ p
n a n a s $ p a n a m a b
n a s $ p a n a m a b a n
s $ p a n a m a b a n a n
```

Adding "a" back to the end of each row should not change the lexicographic ordering of these rows:

```
b a n a n a s $ p a n a m a₁
m a b a n a n a s $ p a n a₂
n a m a b a n a n a s $ p a₃
n a n a s $ p a n a m a b a₄
n a s $ p a n a m a b a n a₅
s $ p a n a m a b a n a n a₆
```

But these are just the rows of M("panamabananas$") containing "a" in *LastColumn*! As a result, the k-th occurrence of "a" in *FirstColumn* corresponds to the k-th occurrence of "a" in *LastColumn*. This argument generalizes for any *symbol* and any string *Text*, which establishes the First-Last property.

Using the First-Last property to invert the Burrows-Wheeler transform

The First-Last Property is interesting, but how can we use it to invert BWT(*Text*) = "ard$rcaaaabb"? Recalling Figure 9.10, let's return to where we were in our attempt to reconstruct the first row of M(*Text*) and index the occurrences of each symbol in *FirstColumn* and *LastColumn*:

```
$₁  a ? ? ? ? ? ? ? ? ? ? a₁
a₁  ? ? ? ? ? ? ? ? ? ? ? r₁
a₂  ? ? ? ? ? ? ? ? ? ? ? d₁
a₃  ? ? ? ? ? ? ? ? ? ? ? $₁
a₄  ? ? ? ? ? ? ? ? ? ? ? r₂
a₅  ? ? ? ? ? ? ? ? ? ? ? c₁
b₁  ? ? ? ? ? ? ? ? ? ? ? a₂
b₂  ? ? ? ? ? ? ? ? ? ? ? a₃
c₁  ? ? ? ? ? ? ? ? ? ? ? a₄
d₁  ? ? ? ? ? ? ? ? ? ? ? a₅
r₁  ? ? ? ? ? ? ? ? ? ? ? b₁
r₂  ? ? ? ? ? ? ? ? ? ? ? b₂
```

The First-Last Property reveals where "a_3" is hiding in *LastColumn*:

HOW DO WE LOCATE DISEASE-CAUSING MUTATIONS?

$$
\begin{array}{llllllllllllll}
\$_1 & \mathbf{a} & ? & ? & ? & ? & ? & ? & ? & ? & ? & a_1 \\
a_1 & ? & ? & ? & ? & ? & ? & ? & ? & ? & ? & r_1 \\
a_2 & ? & ? & ? & ? & ? & ? & ? & ? & ? & ? & d_1 \\
\mathbf{a_3} & ? & ? & ? & ? & ? & ? & ? & ? & ? & ? & \$_1 \\
a_4 & ? & ? & ? & ? & ? & ? & ? & ? & ? & ? & r_2 \\
a_5 & ? & ? & ? & ? & ? & ? & ? & ? & ? & ? & c_1 \\
b_1 & ? & ? & ? & ? & ? & ? & ? & ? & ? & ? & a_2 \\
b_2 & ? & ? & ? & ? & ? & ? & ? & ? & ? & ? & \mathbf{a_3} \\
c_1 & ? & ? & ? & ? & ? & ? & ? & ? & ? & ? & a_4 \\
d_1 & ? & ? & ? & ? & ? & ? & ? & ? & ? & ? & a_5 \\
r_1 & ? & ? & ? & ? & ? & ? & ? & ? & ? & ? & b_1 \\
r_2 & ? & ? & ? & ? & ? & ? & ? & ? & ? & ? & b_2 \\
\end{array}
$$

Since we know that "a_3" is located at the end of the eighth row, we can wrap around this row to determine that "b_2" follows "a_3" in *Text*. Thus, the second symbol of *Text* is "b", which we can now add to the first row of M(*Text*):

$$
\begin{array}{llllllllllllll}
\$_1 & \mathbf{a} & \mathbf{b} & ? & ? & ? & ? & ? & ? & ? & ? & a_1 \\
a_1 & ? & ? & ? & ? & ? & ? & ? & ? & ? & ? & r_1 \\
a_2 & ? & ? & ? & ? & ? & ? & ? & ? & ? & ? & d_1 \\
\mathbf{a_3} & ? & ? & ? & ? & ? & ? & ? & ? & ? & ? & \$_1 \\
a_4 & ? & ? & ? & ? & ? & ? & ? & ? & ? & ? & r_2 \\
a_5 & ? & ? & ? & ? & ? & ? & ? & ? & ? & ? & c_1 \\
b_1 & ? & ? & ? & ? & ? & ? & ? & ? & ? & ? & a_2 \\
\mathbf{b_2} & ? & ? & ? & ? & ? & ? & ? & ? & ? & ? & \mathbf{a_3} \\
c_1 & ? & ? & ? & ? & ? & ? & ? & ? & ? & ? & a_4 \\
d_1 & ? & ? & ? & ? & ? & ? & ? & ? & ? & ? & a_5 \\
r_1 & ? & ? & ? & ? & ? & ? & ? & ? & ? & ? & b_1 \\
r_2 & ? & ? & ? & ? & ? & ? & ? & ? & ? & ? & b_2 \\
\end{array}
$$

In Figure 9.12, we illustrate repeated applications of the First-Last Property to reconstruct more and more symbols from *Text*. Presto — the string that we have been trying to reconstruct is "abracadabra$".

EXERCISE BREAK: Reconstruct the string whose Burrows-Wheeler transform is "enwvpeoseu$llt".

STOP and Think: Can *any* string (having a single "$" symbol) be inverted using the inverse Burrows-Wheeler transform?

You are now ready to implement the inverse of the Burrows-Wheeler transform.

CHAPTER 9

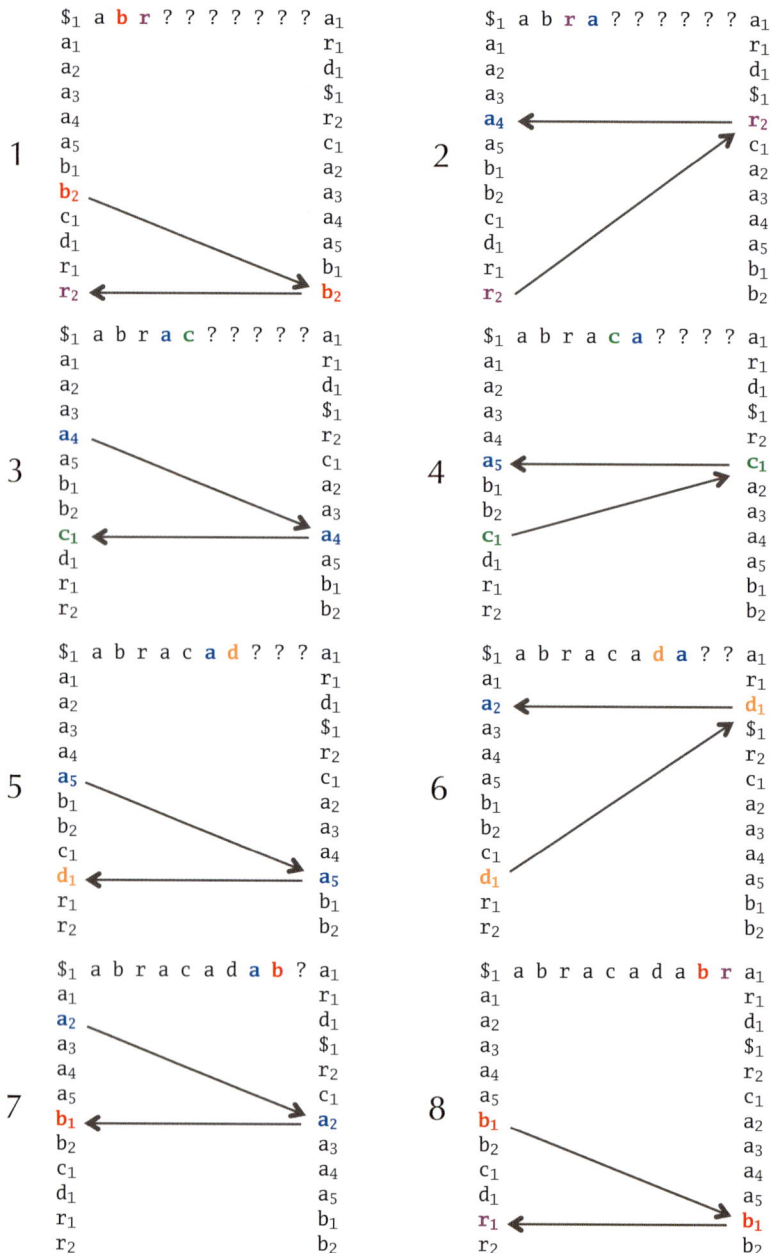

FIGURE 9.12 Repeated applications of the First-Last Property reconstruct the string "abracadabra$" from its Burrows-Wheeler transform "ard$rcaaaabb".

146

How Do We Locate Disease-Causing Mutations?

Inverse Burrows-Wheeler Transform Problem:

Reconstruct a string from its Burrows-Wheeler transform.

Input: A string *Transform* (with a single "$" symbol).
Output: The string *Text* such that BWT(*Text*) = *Transform*.

Pattern Matching with the Burrows-Wheeler Transform

A first attempt at Burrows-Wheeler pattern matching

The Burrows-Wheeler transform may be fascinating, but how could it possibly help us decrease the memory required for pattern matching? The idea motivating a Burrows-Wheeler-based approach to pattern matching relies on the observation that each row of M(*Text*) begins with a different suffix of *Text*. Since these suffixes are already ordered lexicographically, as we already noted when pattern matching with the suffix array, any matches of *Pattern* in *Text* will appear at the beginning of consecutive rows of M(*Text*), as shown in Figure 9.13.

M(*Text*)	SUFFIXARRAY(*Text*)
$ p a n a m a b a n a n a s	13
a b a n a n a s $ p a n a m	5
a m a b a n a n a s $ p a n	3
a n a m a b a n a n a s $ p	1
a n a n a s $ p a n a m a b	7
a n a s $ p a n a m a b a n	9
a s $ p a n a m a b a n a n	11
b a n a n a s $ p a n a m a	6
m a b a n a n a s $ p a n a	4
n a m a b a n a n a s $ p a	2
n a n a s $ p a n a m a b a	8
n a s $ p a n a m a b a n a	10
p a n a m a b a n a n a s $	0
s $ p a n a m a b a n a n a	12

FIGURE 9.13 (Left) Because the rows of M(*Text*) are ordered lexicographically, suffixes beginning with the same string ("**ana**") appear in consecutive rows of the matrix. (Right) The suffix array records the starting position of each suffix in *Text* and immediately tells us the locations of "**ana**".

CHAPTER 9

We now have the outline of a method to match *Pattern* to *Text*. Construct M(*Text*), and then identify rows beginning with the first symbol of *Pattern*. Among these rows, determine which ones have a second element matching the second symbol of *Pattern*. Continue this process until we find which rows of M(*Text*) begin with *Pattern*.

STOP and Think: What is wrong with this approach?

Moving backward through a pattern

The issue with this proposed method for pattern matching is that we cannot afford storing the entire matrix M(*Text*), which has $|Text|^2$ entries. In an effort to reduce memory requirements, let's forbid ourselves from accessing any information in M(*Text*) other than *FirstColumn* and *LastColumn*. Using these two columns, we will try to match *Pattern* to *Text* by moving *backward* through *Pattern*. For example, if we want to match *Pattern* = "ana" to *Text* = "panamabananas$", then we will first identify rows of M(*Text*) beginning with "a", the last letter of "ana":

$\$_1$ p a n a m a b a n a n a s_1
a_1 b a n a n a s $ p a n a m_1
a_2 m a b a n a n a s $ p a n_1
a_3 n a m a b a n a n a s $ p_1
a_4 n a n a s $ p a n a m a b_1
a_5 n a s $ p a n a m a b a n_2
a_6 s $ p a n a m a b a n a n_3
b_1 a n a n a s $ p a n a m a_1
m_1 a b a n a n a s $ p a n a_2
n_1 a m a b a n a n a s $ p a_3
n_2 a n a s $ p a n a m a b a_4
n_3 a s $ p a n a m a b a n a_5
p_1 a n a m a b a n a n a s $\$_1$
s_1 $ p a n a m a b a n a n a_6

As we are moving backward through "ana", we will next look for rows of M(*Text*) beginning with "na". To do this without knowing the entire matrix M(*Text*), we again use the fact that a symbol in *LastColumn* must precede the symbol of *Text* found in the same row in *FirstColumn*. Thus, we only need to identify those rows of M(*Text*) beginning with "a" and ending with "n":

HOW DO WE LOCATE DISEASE-CAUSING MUTATIONS?

```
$₁  p a n a m a b a n a n a  s₁
a₁  b a n a n a s $ p a n a  m₁
a₂  m a b a n a n a s $ p a  n₁
a₃  n a m a b a n a n a s $  p₁
a₄  n a n a s $ p a n a m a  b₁
a₅  n a s $ p a n a m a b a  n₂
a₆  s $ p a n a m a b a n a  n₃
b₁  a n a n a s $ p a n a m  a₁
m₁  a b a n a n a s $ p a n  a₂
n₁  a m a b a n a n a s $ p  a₃
n₂  a n a s $ p a n a m a b  a₄
n₃  a s $ p a n a m a b a n  a₅
p₁  a n a m a b a n a n a s  $₁
s₁  $ p a n a m a b a n a n  a₆
```

The First-Last Property tells us where to find the three highlighted "n" in *FirstColumn*, as shown below. All three rows end with "a", yielding three total occurrences of "ana" in *Text*.

```
$₁  p a n a m a b a n a n a  s₁
a₁  b a n a n a s $ p a n a  m₁
a₂  m a b a n a n a s $ p a  n₁
a₃  n a m a b a n a n a s $  p₁
a₄  n a n a s $ p a n a m a  b₁
a₅  n a s $ p a n a m a b a  n₂
a₆  s $ p a n a m a b a n a  n₃
b₁  a n a n a s $ p a n a m  a₁
m₁  a b a n a n a s $ p a n  a₂
n₁  a m a b a n a n a s $ p  a₃
n₂  a n a s $ p a n a m a b  a₄
n₃  a s $ p a n a m a b a n  a₅
p₁  a n a m a b a n a n a s  $₁
s₁  $ p a n a m a b a n a n  a₆
```

The highlighted occurrences of "a" in *LastColumn* correspond to the third, fourth, and fifth occurrences of "a" in this column, and the First-Last Property tells us that they should correspond to the third, fourth, and fifth occurrences of "a" in *FirstColumn* as well, which identifies the three matches of "ana":

CHAPTER 9

```
$₁  p a n a m a b a n a n a s  s₁
a₁  b a n a n a s $ p a n a m  m₁
a₂  m a b a n a n a s $ p a n  n₁
a₃  n a m a b a n a n a s $ p  p₁
a₄  n a n a s $ p a n a m a b  b₁
a₅  n a s $ p a n a m a b a n  n₂
a₆  s $ p a n a m a b a n a n  n₃
b₁  a n a n a s $ p a n a m a  a₁
m₁  a b a n a n a s $ p a n a  a₂
n₁  a m a b a n a n a s $ p a  a₃
n₂  a n a s $ p a n a m a b a  a₄
n₃  a s $ p a n a m a b a n a  a₅
p₁  a n a m a b a n a n a s $  $₁
s₁  $ p a n a m a b a n a n a  a₆
```

EXERCISE BREAK: Match *Pattern* = "banana" to *Text* = "panamabananas$" by walking backward through *Pattern* using the Burrows-Wheeler transform of *Text*.

The Last-to-First mapping

We now know how to use BWT(*Text*) to find all matches of *Pattern* in *Text* by walking backward through *Pattern*. However, every time we walk backward, we need to keep track of the rows of M(*Text*) where the matches of a suffix of *Pattern* are hiding. Fortunately, we know that at each step, the rows of M(*Text*) that match a suffix of *Pattern* clump together in consecutive rows of M(*Text*). This means that the collection of all matching rows is revealed by only two pointers, *top* and *bottom*: *top* holds the index of the first row of M(*Text*) that matches the current suffix of *Pattern*, and *bottom* holds the index of the last row of M(*Text*) that matches this suffix. Figure 9.14 shows the process of updating pointers; after walking backward through *Pattern* = "ana", we have that *top* = 3 and *bottom* = 5. After traversing *Pattern*, we can compute the total number of matches of *Pattern* in *Text* by calculating *bottom* − *top* + 1 (e.g., there are 5 − 3 + 1 = 3 matches of "ana" in "panamabananas$").

Let's concentrate on how pointers are updated from one stage to the next. Consider the transition from the second to the third panel in Figure 9.14; how did we know to update the pointers (*top* = 1, *bottom* = 6) into (*top* = 9, *bottom* = 11)? We are looking for the first and last occurrence of "n" in the range of positions from *top* = 1 to *bottom* = 6 in *LastColumn*. The first occurrence of "n" in this range is "n_1" (in position 2) and the last is "n_3" (position 6).

In order to update the *top* and *bottom* pointers, we need to determine where "n_1" and "n_3" occur in *FirstColumn*. The **Last-to-First mapping**, denoted LASTTOFIRST(*i*), answers the following question: given a symbol at position *i* in *LastColumn*, what is its position in *FirstColumn*?

For our ongoing example, LASTTOFIRST(2) = 9, since the symbol at position 2 of *LastColumn* ("n_1") occurs at position 9 in *FirstColumn*, as shown in Figure 9.15. Similarly, LASTTOFIRST(6) = 11, since the symbol at position 6 of *LastColumn* ("n_3") occurs at position 11 in *FirstColumn*. Therefore, with the help of the Last-to-First mapping, we can quickly update the pointers (*top* = 1, *bottom* = 6) into (*top* = 9, *bottom* = 11).

We are now ready to describe **BWMATCHING**, an algorithm that counts the total number of matches of *Pattern* in *Text*, where the only information that we are given is *FirstColumn* and *LastColumn* in addition to the Last-to-First mapping. The pointers *top* and *bottom* are updated by the green lines in the following pseudocode.

BWMATCHING(*FirstColumn*, *LastColumn*, *Pattern*, LASTTOFIRST)
    top ← 0
    bottom ← |*LastColumn*| − 1
    while *top* ≤ *bottom*
        if *Pattern* is nonempty
            symbol ← last letter in *Pattern*
            remove last letter from *Pattern*
            if positions from *top* to *bottom* in *LastColumn* contain *symbol*
                topIndex ← first position of *symbol* among positions from *top* to *bottom*
                      in *LastColumn*
                bottomIndex ← last position of *symbol* among positions from *top* to
                      bottom in *LastColumn*
                top ← LASTTOFIRST(*topIndex*)
                bottom ← LASTTOFIRST(*bottomIndex*)
            else
                return 0
        else
            return *bottom* − *top* + 1

CHAPTER 9

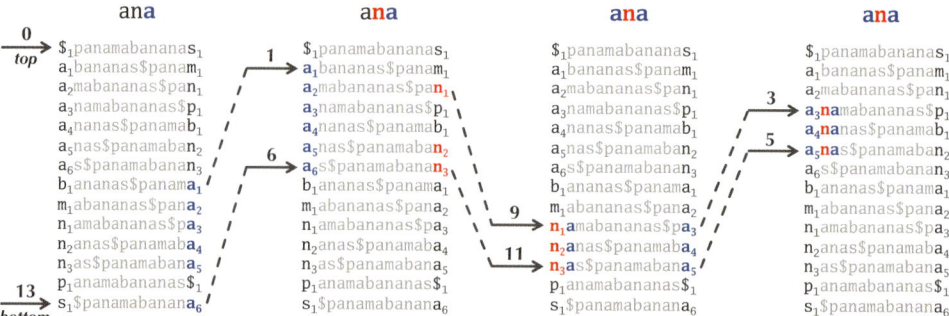

FIGURE 9.14 The pointers *top* and *bottom* hold the indices of the first and last rows of M(*Text*) matching the current suffix of *Pattern* = "ana". The above diagram shows how these pointers are updated when walking backwards through "ana" and looking for substring matches in "panamabananas$".

i	FirstColumn	LastColumn	LastToFirst(i)	COUNT					
				$	a	b	m	n	p	s
0	$_1$	s_1	13	0	0	0	0	0	0	0
1	a_1	m_1	8	0	0	0	0	0	0	1
2	a_2	n_1	9	0	0	0	1	0	0	1
3	a_3	p_1	12	0	0	0	1	1	0	1
4	a_4	b_1	7	0	0	0	1	1	1	1
5	a_5	n_2	10	0	0	1	1	1	1	1
6	a_6	n_3	11	0	0	1	1	2	1	1
7	b_1	a_1	1	0	0	1	1	3	1	1
8	m_1	a_2	2	0	1	1	1	3	1	1
9	n_1	a_3	3	0	2	1	1	3	1	1
10	n_2	a_4	4	0	3	1	1	3	1	1
11	n_3	a_5	5	0	4	1	1	3	1	1
12	p_1	$_1	0	0	5	1	1	3	1	1
13	s_1	a_6	6	1	5	1	1	3	1	1
				1	6	1	1	3	1	1

FIGURE 9.15 The Last-to-First mapping and count array. Precomputing the count array prevents time-consuming updates of the *top* and *bottom* pointers in **BWMatching**.

Speeding Up Burrows-Wheeler Pattern Matching

Substituting the Last-to-First mapping with count arrays

If you implemented **BWMATCHING** in the previous section, you probably found this algorithm to be slow. The reason for its sluggishness is that updating the pointers *top* and *bottom* is time-intensive, since it requires examining every symbol in *LastColumn* between *top* and *bottom* at each step. To improve **BWMATCHING**, we introduce a function $\text{COUNT}_{symbol}(i, LastColumn)$, which returns the number of occurrences of *symbol* in the first i positions of *LastColumn*. For example, $\text{COUNT}_{\text{"n"}}(10, \text{"smnpbnnaaaaa\$a"}) = 3$, and $\text{COUNT}_{\text{"a"}}(4, \text{"smnpbnnaaaaa\$a"}) = 0$. In Figure 9.15, we show arrays holding $\text{COUNT}_{symbol}(i, \text{"smnpbnnaaaaa\$a"})$ for every *symbol* occurring in "panamabananas\$".

EXERCISE BREAK: Compute the arrays COUNT for BWT("abracadabra\$").

We will say that the k-th occurrence of *symbol* in a column of a matrix has **rank** k in this column. For *Text* = "panamabananas\$", note that the first and last occurrences of *symbol* in the range of positions from *top* to *bottom* in *LastColumn* have respective ranks

$$\text{COUNT}_{symbol}(top, LastColumn) + 1$$

and

$$\text{COUNT}_{symbol}(bottom + 1, LastColumn).$$

As illustrated in Figure 9.15, when $top = 1$, $bottom = 6$, and $symbol =$ "n",

$$\text{COUNT}_{\text{"n"}}(top, LastColumn) + 1 = 1$$
$$\text{COUNT}_{\text{"n"}}(bottom + 1, LastColumn) = 3$$

The occurrences of "n" having ranks 1 and 3 are located at positions 2 and 6 of *LastColumn*, implying that we should update *top* to LASTTOFIRST(2) = 9 and *bottom* to LASTTOFIRST(6) = 11. Thus, the four green lines in the pseudocode for **BWMATCHING** can be rewritten as follows.

topIndex ← position of *symbol* with rank $\text{COUNT}_{symbol}(top, LastColumn) + 1$
        in *LastColumn*
bottomIndex ← position of *symbol* with rank $\text{COUNT}_{symbol}(bottom + 1, LastColumn)$
        in *LastColumn*
top ← LASTTOFIRST(*topIndex*)
bottom ← LASTTOFIRST(*bottomIndex*)

By eliminating the variables *topIndex* and *bottomIndex*, we can reduce these four lines of pseudocode to only two lines:

> *top* ← LASTTOFIRST(position of *symbol* with rank COUNT$_{symbol}$(*top*, *LastColumn*) + 1 in *LastColumn*)
> *bottom* ← LASTTOFIRST(position of *symbol* with rank COUNT$_{symbol}$(*bottom* + 1, *LastColumn*) in *LastColumn*)

Note that these two lines of pseudocode merely compute the position of *symbol* with rank *i* in *FirstColumn* from its positions in *LastColumn*. This task can be compactly described without the Last-to-First mapping by the following two lines:

> *top* ← position of *symbol* with rank COUNT$_{symbol}$(*top*, *LastColumn*) + 1 in *FirstColumn*
> *bottom* ← position of *symbol* with rank COUNT$_{symbol}$(*bottom* + 1, *LastColumn*) in *FirstColumn*

For *top* = 1, *bottom* = 6, and *symbol* = "n", the occurrences of "n" having ranks COUNT$_{"n"}$(*top*, *LastColumn*) + 1 = 1 and COUNT$_{"a"}$(*bottom* + 1, *LastColumn*) = 3 are located in positions 9 and 11 of *FirstColumn*, respectively.

STOP and Think: Do we need to store all of *FirstColumn* in memory in order to execute the preceding two lines of pseudocode?

Getting rid of the first column of the Burrows-Wheeler matrix

BWMATCHING requires us to store *FirstColumn*, *LastColumn*, and LASTTOFIRST. We can reduce the amount of memory needed to store the information contained in *FirstColumn* by defining FIRSTOCCURRENCE(*symbol*) as the first position of *symbol* in *FirstColumn*. If *Text* = "panamabananas$", then *FirstColumn* is "$aaaaaabmnnnps", and the array holding all values of FIRSTOCCURRENCE is [0, 1, 7, 8, 9, 11, 12], as shown in Figure 9.16. For DNA strings of any length, the array FIRSTOCCURRENCE contains only five elements. The previous two lines of code can now be rewritten as follows:

> *top* ← FIRSTOCCURRENCE(*symbol*) + COUNT$_{symbol}$(*top*, *LastColumn*)
> *bottom* ← FIRSTOCCURRENCE(*symbol*) + COUNT$_{symbol}$(*bottom* + 1, *LastColumn*) − 1

i	FirstColumn	FirstOccurrence
0	$\$_1$	0
1	a_1	1
2	a_2	
3	a_3	
4	a_4	
5	a_5	
6	a_6	
7	b_1	7
8	m_1	8
9	n_1	9
10	n_2	
11	n_3	
12	p_1	12
13	s_1	13

FIGURE 9.16 The array FirstOccurrence has just seven elements, which is equal to the number of distinct symbols in "panamabananas$".

When $top = 1$, $bottom = 6$, and $symbol = $ "n", we have that

$$\text{FirstOccurrence}(\text{"n"}) = 9$$
$$\text{Count}_{\text{"n"}}(top, LastColumn) = 0$$
$$\text{Count}_{\text{"n"}}(bottom + 1, LastColumn) = 3$$

Recalling Figure 9.14, this again implies that $(top = 1, bottom = 6)$ will be updated as

$$top = 9 + 0 = 9$$
$$bottom = 9 + 3 - 1 = 11$$

In the process of simplifying the green lines of pseudocode from **BWMatching**, we have also substituted *FirstColumn* by FirstOccurrence and LastToFirst by Count, resulting in a more efficient algorithm called **BetterBWMatching**, shown below.

You may be wondering why we call this algorithm "better", since on the one hand, if you have to compute the Count arrays as you go, then you will not obtain a runtime speedup. On the other hand, if you have to precompute these arrays, then you will need to store them in memory, which is space-intensive. Hold onto this thought.

CHAPTER 9

> **BETTERBWMATCHING**(FIRSTOCCURRENCE, *LastColumn*, *Pattern*, COUNT)
> $top \leftarrow 0$
> $bottom \leftarrow |LastColumn| - 1$
>   **while** $top \leq bottom$
>   **if** *Pattern* is nonempty
> $symbol \leftarrow$ last letter in *Pattern*
>   remove last letter from *Pattern*
> $top \leftarrow \text{FIRSTOCCURRENCE}(symbol) + \text{COUNT}_{symbol}(top, LastColumn)$
> $bottom \leftarrow \text{FIRSTOCCURRENCE}(symbol) + \text{COUNT}_{symbol}(bottom + 1,$
> $LastColumn) - 1$
>   **else**
>   **return** $bottom - top + 1$
>   **return**

Where are the Matched Patterns?

We hope that you have noticed a limitation of **BETTERBWMATCHING** — even though this algorithm counts the *number* of occurrences of *Pattern* in *Text*, it does not tell us *where* these occurrences are located in *Text*! To locate pattern matches identified by the algorithm, we can once again use the suffix array, as shown in Figure 9.13 (right). In this figure, the suffix array immediately finds the three matches of "ana" in "panamabananas$".

The suffix array makes our job easy, but recall that our original motivation for using the Burrows-Wheeler transform was to *reduce* the amount of memory used by the suffix array for pattern matching. If we add the suffix array to Burrows-Wheeler-based pattern matching, then we are right back where we started!

The memory-saving device that we will employ is inelegant but useful. We will build a **partial suffix array** of *Text*, denoted $\text{SUFFIXARRAY}_K(Text)$, which only contains values that are multiples of some positive integer K (Figure 9.17). In real applications, partial suffix arrays are often constructed for $K = 100$, thus reducing memory usage by a factor of 100 compared to a full suffix array. See **CHARGING STATION: Partial Suffix Array Construction** for more details.

HOW DO WE LOCATE DISEASE-CAUSING MUTATIONS?

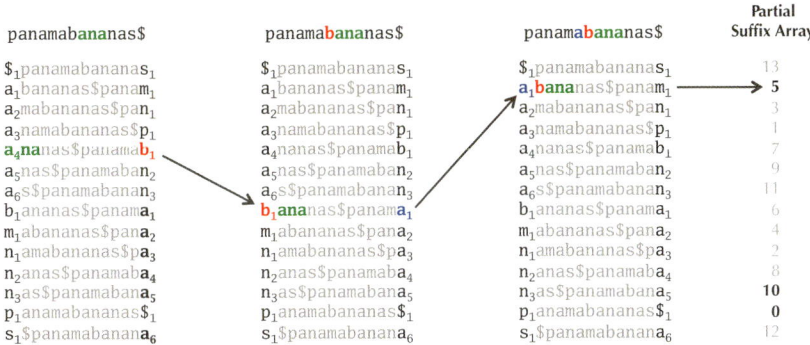

FIGURE 9.17 One of the matches of "ana" in "panama**b**ana**na**s$" is highlighted in the matrix on the right. By walking backward, we find that "**ana**" is preceded by "**b$_1$**", which in turn is preceded by "**a$_1$**". The partial suffix array above, generated for $K = 5$, indicates that "a$_1$" occurs at position 5 of "panama**b**ana**na**s". Since it took us two steps to walk backward to "a$_1$", we conclude that this occurrence of "ana" begins at position $5 + 2 = 7$.

Burrows and Wheeler Set Up Checkpoints

We will now discuss how to improve **BETTERBWMATCHING** by resolving the trade-off between precomputing the values of COUNT$_{symbol}(i, LastColumn)$ (requiring substantial memory) and computing these values as we go (requiring substantial runtime).

The balance that we strike is similar to the one used for the partial suffix array. Rather than storing COUNT$_{symbol}(i, LastColumn)$ for *all* positions *i*, we will only store the COUNT arrays when *i* is divisible by *C*, where *C* is a constant; these arrays are called **checkpoint arrays** (Figure 9.18). When *C* is large (*C* is typically equal to 100 in practice) and the alphabet is small (e.g., four nucleotides), checkpoint arrays require only a fraction of the memory used by BWT(*Text*).

What about runtime? Using checkpoint arrays, we can compute the *top* and *bottom* pointers in a constant number of steps (i.e., fewer than *C*). Since each string *Pattern* requires at most |*Pattern*| pointer updates, the modified **BETTERBWMATCHING** algorithm now requires $\mathcal{O}(|Patterns|)$ runtime, which is the same as using a trie or suffix array.

Furthermore, we now only have to store the following data in memory: BWT(*Text*), FIRSTOCCURRENCE, the partial suffix array, and the checkpoint arrays. Storing this data requires memory approximately equal to $1.5 \cdot |Text|$. Thus, we have finally knocked

i	LastColumn	Count						
		$	a	b	m	n	p	s
0	s_1	**0**	**0**	**0**	**0**	**0**	**0**	**0**
1	m_1	0	0	0	0	0	0	1
2	n_1	0	0	0	1	0	0	1
3	p_1	0	0	0	1	1	0	1
4	b_1	0	0	0	1	1	1	1
5	n_2	**0**	**0**	**1**	**1**	**1**	**1**	**1**
6	n_3	0	0	1	1	2	1	1
7	a_1	0	0	1	1	3	1	1
8	a_2	0	1	1	1	3	1	1
9	a_3	0	2	1	1	3	1	1
10	a_4	**0**	**3**	**1**	**1**	**3**	**1**	**1**
11	a_5	0	4	1	1	3	1	1
12	$\$_1$	0	5	1	1	3	1	1
13	a_6	1	5	1	1	3	1	1
		1	6	1	1	3	1	1

FIGURE 9.18 The Count checkpoint arrays for *Text* = "panamabananas\$" and $C = 5$ are highlighted in bold. If we want to compute Count"a"(13, "smnpbnnaaaaa\$a"), then the checkpoint array at position 10 tells us that there are 3 occurrences of "a" *before* position 10 of "smnpbnn**aaa**aa\$a". We then check whether "a" is present at position **10** (yes), **11** (yes), and **12** (no) of *LastColumn* to conclude that Count"a"(13, "smnpbnnaaaaa\$a") $= 3 + 2 = 5$.

down the memory required for solving the Multiple Pattern Matching Problem for millions of sequencing reads into a workable range.

As the plummeting cost of DNA sequencing has made headlines, it is easy to fail to appreciate the progress that has been made in the computational side of read mapping. So, before continuing into approximate pattern matching, we would like to pause and reflect on just how far the algorithms for read mapping – and the hardware running them – have progressed. In 1975, the state-of-the-art Aho-Corasick algorithm was touted as requiring 15 minutes to map just 24 English words to a dictionary. Just a generation later, when Burrows-Wheeler based approaches were introduced into read mapping, the same 15 minutes was used to map almost *ten million* reads to a reference genome. Having seen how far we have come in the last four decades, we can only imagine where it will take us in the future.

HOW DO WE LOCATE DISEASE-CAUSING MUTATIONS?

Epilogue: Mismatch-Tolerant Read Mapping

Reducing approximate pattern matching to exact pattern matching

In this section, we will return to the goal of identifying SNPs in an individual genome when compared against the reference genome. To do so, we need to generalize the Approximate Pattern Matching Problem from Chapter 1 to the case of multiple patterns.

Multiple Approximate Pattern Matching Problem:
Find all approximate occurrences of a collection of patterns in a text.

Input: A string *Text*, a collection of shorter strings *Patterns*, and an integer d.
Output: All starting positions in *Text* where a string from *Patterns* appears as a substring with at most d mismatches.

We begin with the simple observation that if *Pattern* occurs in *Text* with a single mismatch, then we can divide *Pattern* into two halves, one of which occurs exactly in *Text*, as illustrated below.

> *Pattern*   acttggct
> *Text*   ...ggcacactaggctcc...

Thus, we can find whether *Pattern* occurs with one mismatch in *Text* by dividing *Pattern* into two halves and then searching for exact matches of these shorter strings. If we find a match with one of these halves of *Pattern*, we then check if the entire string *Pattern* occurs with a single mutation.

This method can easily be generalized for approximate pattern matching with $d > 1$ mismatches; if *Pattern* approximately matches a substring of *Text* with at most d mismatches, then *Pattern* and *Text* must share at least one k-mer for a sufficiently large value of k. For example, if we are looking for a pattern of length 20 with at most $d = 3$ mismatches, then we can divide this pattern into four parts of length $20/(3+1) = 5$ and search for exact matches of these shorter substrings:

> *Pattern*   acttaggctcgggataatcc
> *Text*   ...actaagtctcgggataagcc...

CHAPTER 9

This observation is helpful because it reduces *approximate* pattern matching to the *exact* matching of shorter patterns, which allows us to use fast algorithms that are designed for exact pattern matching but are not applicable to approximate pattern matching, such as approaches based on suffix trees and suffix arrays.

> **STOP and Think:** If *Pattern* has length 23 and appears in *Text* with 3 mismatches, can we conclude that *Pattern* shares a 6-mer with *Text*? Can we conclude that it shares a 5-mer with *Text*?

The question remains how to find the maximum value of k that guarantees that any *Pattern* of length n with d mismatches in *Text* will have a k-mer substring exactly matching *Text*.

Theorem. *If two strings of length n match with at most d mismatches, then they must share a k-mer of length $k = \lfloor n/(d+1) \rfloor$.*

Proof. Divide the first string into $d + 1$ substrings, where the first d substrings have exactly k symbols and the final substring has at least k symbols. For the case $d = 3$, here is a subdivision of a 23-nucleotide string ($n = 23$) into $d + 1 = 4$ substrings, where the first 3 substrings have $k = \lfloor n/(d+1) \rfloor = \lfloor 23/4 \rfloor = 5$ symbols and the last substring has 8 symbols.

acttaggctc**gggat**aatccgga

If you distribute d mismatches among the positions of the string, then the mismatches may affect at most d substrings, which leaves at least one substring (of length at least k) unchanged. This substring is shared by both strings. □

We now have the outline of an algorithm for matching a string *Pattern* of length n to *Text* with at most d mismatches. We first divide *Pattern* into $d + 1$ segments of length $k = \lfloor n/(d+1) \rfloor$, called **seeds**. After finding which seeds match *Text* exactly (**seed detection**), we attempt to extend seeds in both directions in order to verify whether *Pattern* occurs in *Text* with at most d mismatches (**seed extension**).

BLAST: Comparing a sequence against a database

Using shared k-mers to find similarities between biological sequences has some disadvantages. For example, two proteins may have similar functions but not share any k-mers, even for small values of k.

The **Basic Local Alignment Search Tool (BLAST)** is a heuristic that can find similarities between proteins, even if all amino acids in one protein have mutated compared to the other protein. BLAST is so fast that it is often used to query a protein against *all other* known proteins in protein databases. The paper introducing BLAST was released in 1990, and with over 40,000 citations, it has become one of the most cited scientific papers ever published.

To see how BLAST works, say we are given an integer k and strings $x = x_1 \ldots x_n$ and $y = y_1 \ldots y_m$ to compare. We define a **segment pair** of x and y as a pair formed by a k-mer from x and a k-mer from y. The score of the segment pair corresponding to the k-mers starting at position i in x and position j in y is

$$\sum_{t=0}^{k-1} \text{SCORE}(x_{i+t}, y_{j+t}),$$

where $\text{SCORE}(x_{i+t}, y_{j+t})$ is determined by a scoring matrix such as the PAM scoring matrix that we introduced in Chapter 5. A **locally maximal segment pair** is a segment pair whose score cannot be increased by extending or shortening both strings in the segment pair. BLAST attempts to find not the highest-scoring segment pair of x and y, but rather all locally maximal segment pairs in these strings with scores above some threshold.

The key ingredient of BLAST is to first quickly find all k-mers that have scores above a given threshold when scored against some k-mer in the query string. If the score threshold is high, then the set of all resulting segment pairs formed between the query string and strings in the database is not too large. In this case, the database can be searched for exact occurrences of these high-scoring k-mers from this set, producing an initial set of seeds. This is an instance of the Multiple Pattern Matching Problem, which we have learned how to solve quickly.

> **EXERCISE BREAK:** Given the PAM$_{250}$ scoring matrix, only five 3-mers score higher than 23 against CFC: CIC, CLC, CMC, CWC, and CYC. How many 3-mers score higher than 20 against CFC?

After finding seeds, BLAST attempts to extend these seeds (allowing for insertions and deletions) in order to obtain locally maximal segment pairs.

> **EXERCISE BREAK:** Given the PAM$_{250}$ scoring matrix, an amino acid k-mer *Peptide*, and a threshold θ, develop an efficient algorithm for finding the exact number of k-mers scoring more than θ against *Peptide*.

CHAPTER 9

Approximate pattern matching with the Burrows-Wheeler transform

To extend the Burrows-Wheeler approach to approximate pattern matching, we will not stop when we encounter a mismatch. On the contrary, we will proceed onward until we either find an approximate match or exceed the limit of d mismatches.

Figure 9.19 illustrates the search for "asa" in "panamabananas$" with at most 1 mismatch. Let's first proceed as in the case of exact pattern matching, threading "asa" backwards using the Burrows-Wheeler transform. After finding six occurrences of "a", we identify six inexact occurrences of "sa": "**p**a", "**m**a", "**b**a", and three occurrences of "**n**a". We note that these three strings have accumulated a mismatch and then continue with all six strings.

In the next step, five of the inexact occurrences of "sa" can be extended into inexact occurrences of "asa" with only a single mismatch: "a**m**a", "a**b**a", and three occurrences of "a**n**a". We fail to extend "**p**a", which we eliminate from consideration.

In practice, this heuristic faces complications. We do not want to start allowing mismatched strings at the early stages of **BETTERBWMATCHING**, or else we will have to consider too many frivolous candidate strings. We may therefore require that a suffix of *Pattern* of some threshold length matches *Text* exactly. Moreover, the method becomes time-intensive when using large values of d, as we must explore many inexact matches. Practical applications often limit the value of d to at most 3.

You should now be ready to design your own approach to solve the Multiple Approximate Pattern Matching Problem and use this solution to map real sequencing reads.

> **CHALLENGE PROBLEM:** Given the Burrows-Wheeler transform and a partial suffix array of the bacterial genome *Mycoplasma pneumoniae* along with a collection of reads, find all reads occurring in the genome with at most one mismatch.

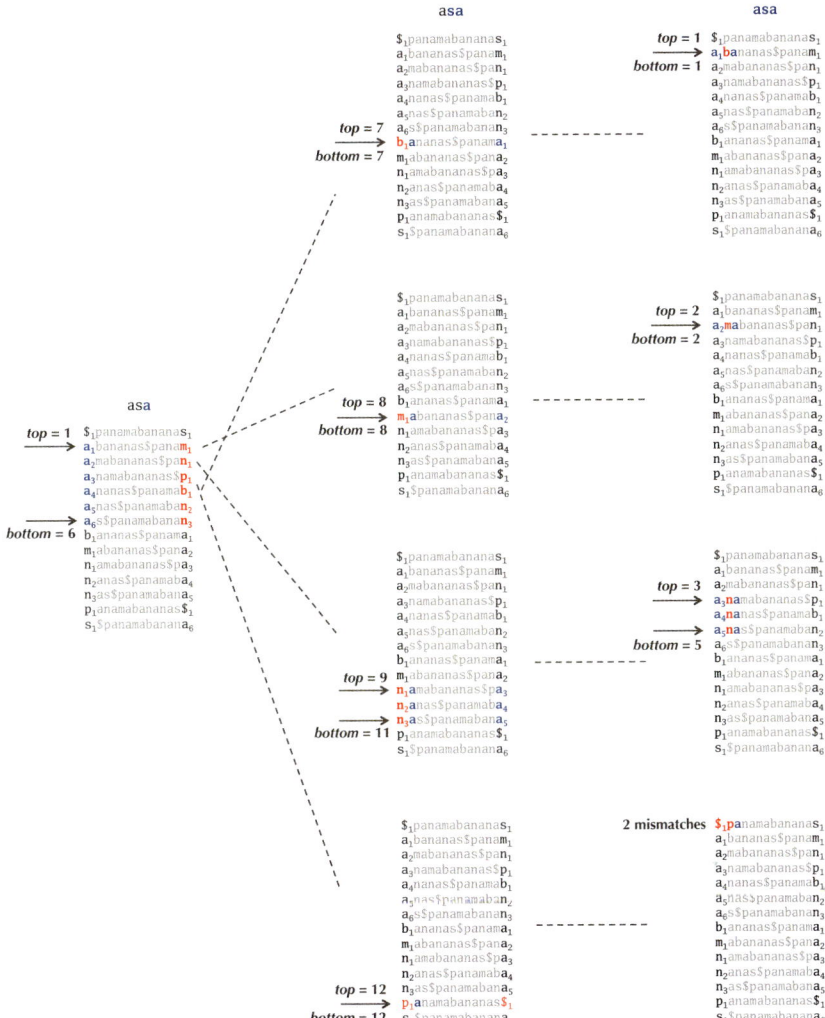

FIGURE 9.19 Using the Burrows-Wheeler transform to find approximate pattern matches of "asa" in "panamabananas$" with 1 mismatch. (Left) We identify six occurrences of "a". (Middle) Working backwards, we find four different inexact partial matches: "ba", "ma", "na" (three occurrences), and "pa". (Right) Additional mismatches are not allowed in the partial matches that we have found, but these partial matches can be extended to yield five inexact matches of "asa".

CHAPTER 9

Charging Stations

Constructing a suffix tree

To construct a suffix tree, we will first modify the construction of the suffix trie as follows. Although each edge *edge* in a suffix trie is labeled by a single symbol SYMBOL(*edge*), it is unclear where this symbol came from in *Text*. We will therefore add another label for each edge (denoted POSITION(*edge*)) referring to the position of this symbol in *Text*. If SYMBOL(*edge*) corresponds to more than one position in *Text*, then we will assign it its minimum starting position.

For example, consider the modified suffix trie for *Text* = "panamabananas$" shown in Figure 9.20. There are five edges labeled by "m" (colored purple), all of which are

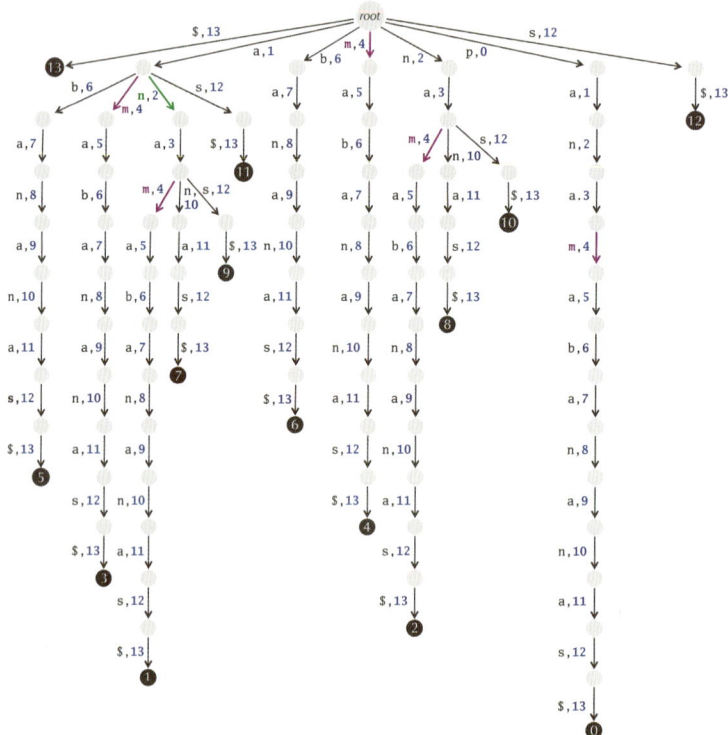

FIGURE 9.20 The modified suffix trie of *Text* = "panamabananas$". Each edge is assigned the minimum position to which the edge's symbol corresponds in *Text* (shown in blue).

labeled by position 4, because this is the only occurrence of "m" in *Text*. On the other hand, the green edge corresponding to "n" is a different story. By following every path from this edge down to the leaves, we see that it corresponds to occurrences of "n" in the suffixes "anamabananas$", "ananas$", and "anas$", at positions 2, 8, and 10. As a result, we assign this edge the minimum of these positions.

The following pseudocode constructs the modified suffix trie of a string *Text* by traversing the suffixes of *Text* from longest to shortest. Given a suffix, it attempts to spell the suffix by moving downward in the tree, following edge labels as far as possible until it can go no further. At that point, it adds the rest of the suffix to the trie in the form of a path to a leaf, along with the position of each symbol in the suffix.

MODIFIEDSUFFIXTRIECONSTRUCTION(*Text*)
    Trie ← a graph consisting of a single node *root*
    for *i* ← 0 to |*Text*| − 1
        currentNode ← *root*
        for *j* ← *i* to |*Text*| − 1
            currentSymbol ← *j*-th symbol of *Text*
            if there is an outgoing edge from *currentNode* labeled by *currentSymbol*
                currentNode ← ending node of this edge
            else
                add a new node *newNode* to *Trie*
                add an edge *newEdge* connecting *currentNode* to *newNode* in *Trie*
                SYMBOL(*newEdge*) ← *currentSymbol*
                POSITION(*newEdge*) ← *j*
                currentNode ← *newNode*
        if *currentNode* is a leaf in *Trie*
            assign label *i* to this leaf
    return *Trie*

We can now transform a modified suffix trie into a suffix tree as follows. Note in Figure 9.6 (page 131) that each edge *edge* in SUFFIXTREE("panamabananas$") is labeled by a string of symbols, denoted STRING(*edge*). As we mentioned in the main text, storing all these strings is memory-intensive, and so we will instead label *edge* by two integers: the starting position of the first occurrence of STRING(*edge*) in *Text*, denoted POSITION(*edge*), and its length, denoted LENGTH(*edge*). For the modified suffix tree of *Text* = "panamabananas$" shown in Figure 9.21, these two integers are colored blue and red, respectively. For example, the edge labeled "mabananas$" in Figure 9.6 is labeled by POSITION(*edge*) = **4** and LENGTH(*edge*) = **10** in Figure 9.21. There are two edges labeled "na" in Figure 9.6, and both of them are labeled by POSITION(*edge*) = **2** and LENGTH(*edge*) = **2** in Figure 9.21.

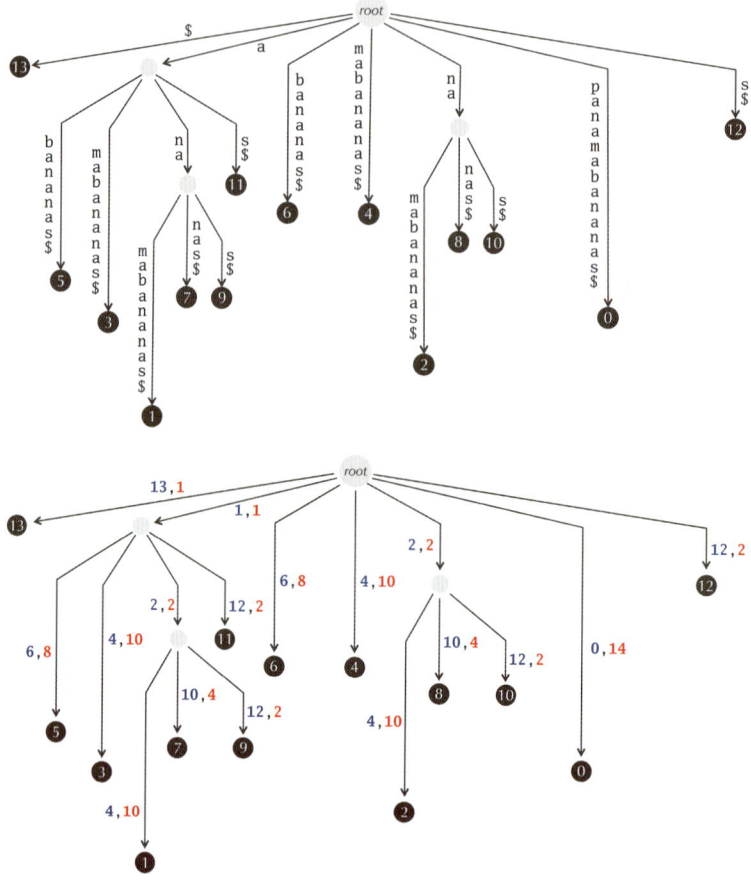

FIGURE 9.21 (Top) The suffix tree of *Text* = "panamabananas$. (Bottom) The modified suffix tree of *Text*. For each edge, the initial position of the substring to which it corresponds in *Text* is shown in blue, and the length of this substring is shown in red.

The following pseudocode constructs a suffix tree using the modified suffix trie constructed by **MODIFIEDSUFFIXTREECONSTRUCTION**. This algorithm will consolidate each non-branching path (i.e., a path whose intermediate nodes have indegree and

How Do We Locate Disease-Causing Mutations?

outdegree equal to 1) of the modified suffix trie into a single edge.

    SUFFIXTREECONSTRUCTION(Text)
        Trie ← MODIFIEDSUFFIXTRIECONSTRUCTION(Text)
        for each non-branching path Path in Trie
            substitute Path by a single edge e connecting the first and last nodes of Path
            POSITION(e) ← POSITION(first edge of Path)
            LENGTH(e) ← number of edges of Path
        return Trie

Solving the Longest Shared Substring Problem

A naive approach for finding a longest shared substring of strings $Text_1$ and $Text_2$ would construct one suffix tree for $Text_1$ and another for $Text_2$. Instead, we will add "#" to the end of $Text_1$, add "$" to the end of $Text_2$, and then construct the single suffix tree for the concatenation of $Text_1$ and $Text_2$ (Figure 9.22). We color a leaf in this suffix tree blue if it is labeled by the starting position of a suffix starting in $Text_1$; we color a leaf red if it is labeled by the starting position of a suffix starting in $Text_2$.

We also color the remaining nodes of the suffix tree blue, red, and purple according to the following rules:

- a node is colored blue or red if all leaves in its subtree (i.e., the subtree beneath it) are all blue or all red, respectively;

- a node is colored purple if its subtree contains both blue and red leaves.

We use COLOR(v) to denote the color of node v.

There are three purple nodes in Figure 9.22 (other than the root), and the strings spelled from the root to each of these nodes are "a", "ana", and "na". Note that these three substrings are shared by $Text_1 =$ "panama" and $Text_2 =$ "bananas". This is no accident.

EXERCISE BREAK: Prove that a path ending in a purple node in the suffix tree of $Text_1$ and $Text_2$ spells out a substring shared by $Text_1$ and $Text_2$.

EXERCISE BREAK: Prove that a path ending in a blue (respectively, red) node in the suffix tree of $Text_1$ and $Text_2$ spells out a substring that appears in $Text_1$ but not in $Text_2$ (respectively, $Text_2$ but not in $Text_1$).

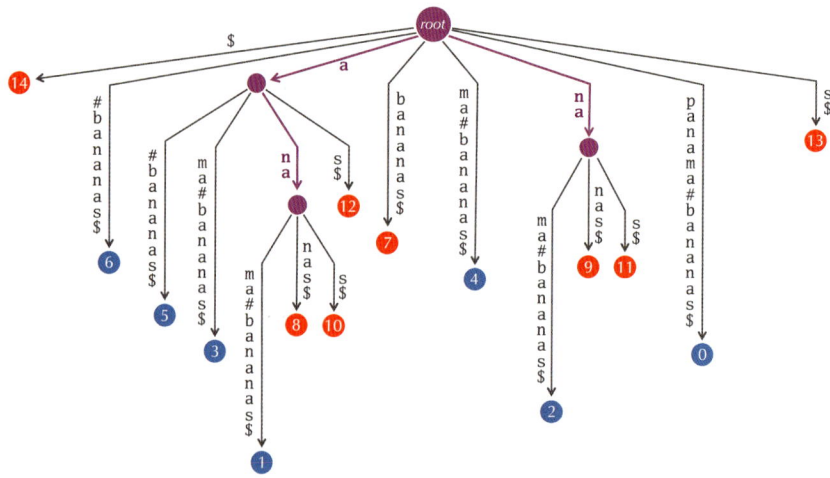

FIGURE 9.22 SUFFIXTREE("panama#bananas$"), constructed for $Text_1$ = "panama" and $Text_2$ = "bananas". Leaves corresponding to suffixes starting in "panama#" are colored blue; leaves corresponding to suffixes starting in "bananas$" are colored red. Every string of symbols spelled from the root to a purple node corresponds to a substring shared by $Text_1$ and $Text_2$.

The previous two exercises imply that in order to find the longest shared substring between $Text_1$ and $Text_2$, we need to examine all purple nodes as well as the strings spelled by paths leading to the purple nodes. A longest such string yields a solution to the Longest Shared Substring Problem.

TREECOLORING, which is illustrated in Figure 9.23, colors the nodes of a suffix tree from the leaves upward. This algorithm assumes that the leaves of the suffix tree have been labeled "blue" or "red" and all other nodes have been labeled "gray". A node in a tree is called **ripe** if it is gray but has no gray children.

```
TREECOLORING(ColoredTree)
    while ColoredTree has ripe nodes
        for each ripe node v in ColoredTree
            if there exist differently colored children of v
                COLOR(v) ← "purple"
            else
                COLOR(v) ← color of all children of v
    return ColoredTree
```

HOW DO WE LOCATE DISEASE-CAUSING MUTATIONS?

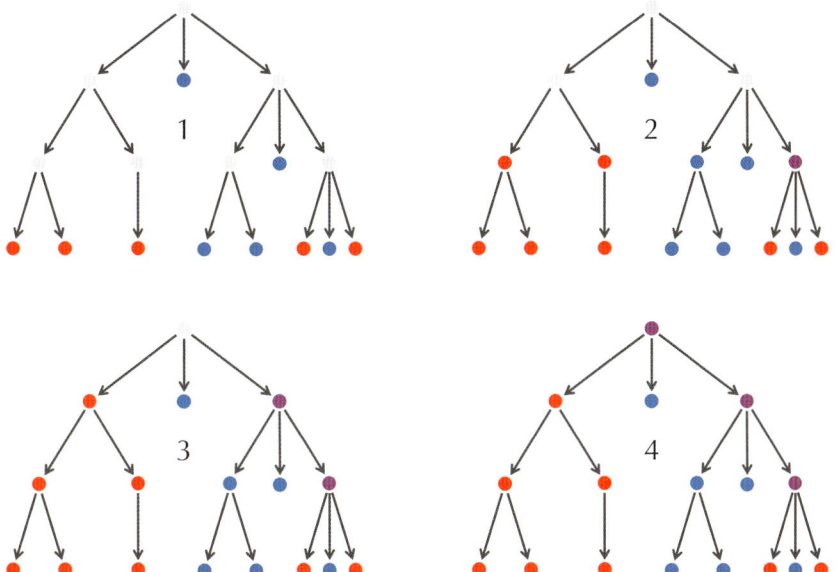

FIGURE 9.23 An illustration of the steps taken by **TREECOLORING** for a tree with an initial coloring of leaves as red or blue (top left).

Partial suffix array construction

To construct the partial suffix array $\text{SUFFIXARRAY}_K(\textit{Text})$, we first need to construct the full suffix array and then retain only the elements of this array that are divisible by K, along with their indices i. This is illustrated in Figure 9.24 for $\textit{Text} = $ "panamabananas\$" and $K = 5$, where $\text{SUFFIXARRAY}_K(\textit{Text})$ corresponds to the bold elements.

i	0	**1**	2	3	4	5	6	7	8	9	10	**11**	**12**	13
SUFFIXARRAY(*Text*)	13	**5**	3	1	7	9	11	6	4	3	8	**10**	**0**	12

FIGURE 9.24 Partial suffix array construction.

CHAPTER 9

Detours

The reference human genome

When genomes are assembled from DNA taken from a number of donors, the reference genome represents a mosaic of donor genomes. The existing reference human genome was derived from thirteen volunteers in the United States; it continues to be improved by fixing errors and filling in the remaining gaps (there are currently over a hundred).

The reference genome is often used as a template on which new individual genomes can be rapidly assembled. Comparison of the reference genome and individual human genomes typically reveals about three million SNPs, and about 0.1% of an individual human genome cannot be matched to the reference genome at all.

In regions with high diversity, the reference genome may differ significantly from individual genomes. An example of a high-diversity region of the human genome is the **major histocompatibility complex**, a family of genes powering immune systems. Since these genes have an unusually large number of alternate forms, two individuals hardly ever have exactly the same set of genes from this complex.

Rearrangements, insertions, and deletions in human genomes

Until recently, biologists focused primarily on SNPs in the human genome, assuming that rearrangements and indels are relatively rare. In 2005, Evan Eichler surprised biologists when he found hundreds of rearrangements and indels separating the genomes of two individuals. This finding was important because rearrangements and indels are often hallmarks of disease; for example, repeated insertions of the nucleotide triplet CAG increases the severity of Huntington's disease.

In 2013, Gerton Lunter revealed the true extent of indels in the human population by identifying over a *million* indels in a cohort of over a hundred individuals. Intriguingly, he found that over half of the indels occurred in just 4% of the genome; in other words, some regions of the human genome represent "indel hotspots". As the catalog of human rearrangements and indels grows, biologists are gaining the ability to identify frequently mutated genes as well as implicate rearrangements and indels when diagnosing complex disorders.

The Aho-Corasick algorithm

The **Aho-Corasick algorithm** for the Multiple Pattern Matching Problem was invented by Alfred Aho and Margaret Corasick in 1975. The runtime of their algorithm is $\mathcal{O}(|Patterns| + |Text| + m)$, where m is the number of output matches.

HOW DO WE LOCATE DISEASE-CAUSING MUTATIONS?

Imagine sliding the trie in Figure 9.25 against *Text* = "bantenna". Similarly to **TRIEMATCHING**, the Aho-Corasick algorithm starts at the root and attempts to build a path spelling a prefix of "bantenna". This attempt fails after three nodes ("**ban**tenna"). In **TRIEMATCHING**, we would begin again at the root and attempt to find a match starting at the second symbol of "bantenna".

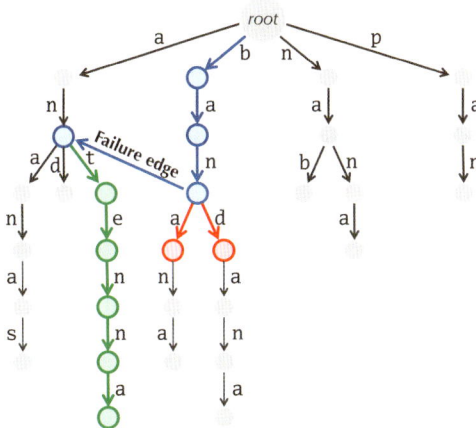

FIGURE 9.25 The trie from Figure 9.1 with an additional failure edge. The failure edge allows us to jump from the path beginning with "ban" to the path beginning with "an".

However, we have thrown away some important information: we *already matched* the first two symbols of "**an**tenna" when searching through "**ban**tenna". Thus, a more sensible strategy is to jump directly to the node "an" and then continue downward in the trie. Indeed, this strategy will eventually match the pattern "**an**tenna". We can implement this "jumping ahead" strategy by augmenting the tree with a **failure edge** connecting node "ban" to node "an" (Figure 9.25). More generally, failure edges are formed by connecting node *v* to node *w* if *w* is the longest suffix of *v* that appears in the trie. The Aho-Corasick algorithm follows failure edges whenever a mismatch is found in order to avoid going all the way back to the root, thus saving time.

From suffix trees to suffix arrays

We can construct SUFFIXARRAY(*Text*) from SUFFIXTREE(*Text*) by applying a **preorder traversal** of SUFFIXTREE(*Text*) as long as the outgoing edges from every node of the suffix tree are arranged lexicographically.

Chapter 9

Given a rooted tree, a node w is called a **child** of a node v if there is an edge connecting v to w in the tree. The preorder traversal of a tree involves visiting a node of the tree, starting at the root, and then recursively preorder traversing the subtrees rooted at each of its children from left to right (Figure 9.26), which is accomplished by the following pseudocode. By taking the order of leaves visited in a preorder traversal of the suffix tree, we obtain the suffix array (consult Figure 9.6 and Figure 9.7).

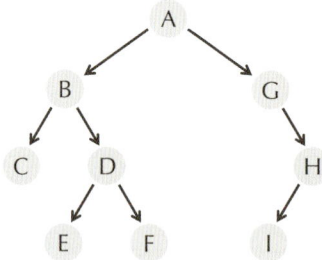

FIGURE 9.26 The preorder traversal of the above tree visits its nodes in increasing order of their labels.

PREORDER(*Tree, Node*)
   visit *Node*
   for each child *Node'* of *Node* from left to right
      PREORDER(*Tree, Node'*)

Conversely, SUFFIXTREE(*Text*) can be constructed from SUFFIXARRAY(*Text*) in linear time by using the **longest common prefix (LCP) array** of *Text*, LCP(*Text*), which stores the length of the longest common prefix shared by consecutive lexicographically ordered suffixes of *Text*. For example, LCP("panamabananas$") is $(0, 0, 1, 1, 3, 3, 1, 0, 0, 0, 2, 2, 0, 0)$, as shown in Figure 9.27.

Suffix Tree Construction from Suffix Array Problem:

Construct a suffix tree from the suffix array and LCP array of a string.

     Input: A string *Text*, its suffix array, and its LCP array.
     Output: The suffix tree of *Text*.

LCP Array	Sorted Suffixes
0	$
0	abananas$
1	amabananas$
1	anamabananas$
3	ananas$
3	anas$
1	as$
0	bananas$
0	mabananas$
0	namabananas$
2	nanas$
2	nas$
0	panamabananas$
0	s$

FIGURE 9.27 The LCP array of "panamabananas$" is formed by sorting the suffixes of "panamabananas$" lexicographically and then finding the length of the longest common prefix shared by consecutive suffixes according to the lexicographic order.

From suffix arrays to suffix trees

Given the suffix array SUFFIXARRAY and LCP array LCP of a string *Text*, the suffix tree SUFFIXTREE(*Text*) can be constructed in linear time using the algorithm illustrated in Figure 9.28. After constructing a **partial suffix tree** for the i lexicographically smallest suffixes (denoted SUFFIXTREE$_i$(*Text*)), this algorithm iteratively inserts the $(i+1)$-th suffix into this tree to form SUFFIXTREE$_{i+1}$(*Text*).

We define the **descent** of a node v in a suffix tree, denoted DESCENT(v), as the length of the concatenation of all path labels from the root to this node. We assume that the descents of all nodes in the growing partial suffix tree have been precomputed.

We start with SUFFIXTREE$_0$(*Text*), which we define as the tree consisting only of the root. To insert the $(i+1)$-th suffix (corresponding to the element SUFFIXARRAY($i+1$) in the suffix array of *Text*) into SUFFIXTREE$_i$(*Text*), we need to know where the path representing this suffix "splits" from the already constructed partial suffix tree

CHAPTER 9

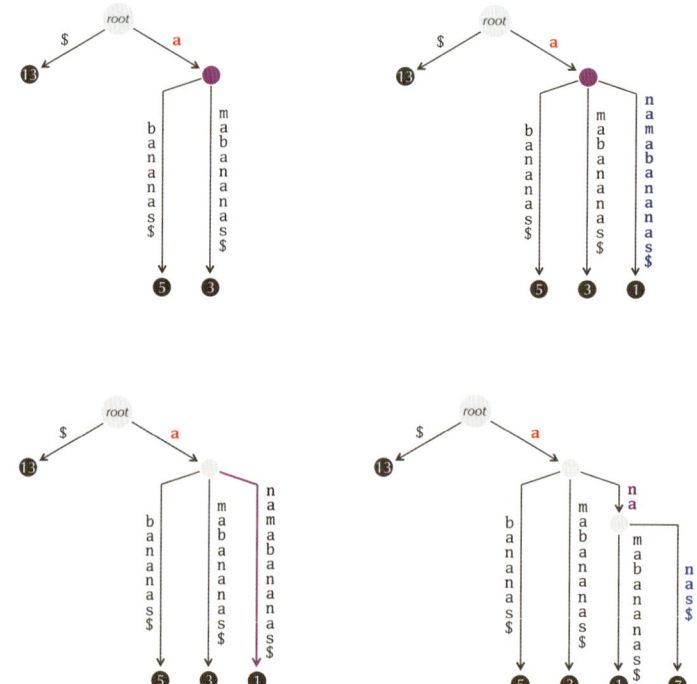

FIGURE 9.28 (Top left) SUFFIXTREE$_3$(*Text*) for *Text* = "panamabananas$". The fourth lexicographically ordered suffix of *Text* is "**anamabananas$**", and we can thread only its first letter into this suffix tree. Our stopping point is at the purple node, so we create a new node branching off this node with edge labeled "**namabananas$**" to obtain SUFFIXTREE$_4$(*Text*) (top right). (Bottom left) The fifth lexicographically ordered suffix of *Text* is "**ananas$**", the first three symbols of which we can thread into SUFFIXTREE$_4$(*Text*) until we reach a stopping point in the middle of the edge "namabananas$". (Bottom right) To form SUFFIXTREE$_5$(*Text*), we create a new node in the middle of the purple edge and branch from this node into two edges: one labeled "**mabananas$**" in order to retain the suffix "anamabananas$" and another labeled "**nas$**" to spell the new suffix "ananas$". In general, the number of red symbols that we spell down in SUFFIXTREE$_i$(*Text*) to form SUFFIXTREE$_{i+1}$(*Text*) is given by the $(i + 1)$-th entry in the LCP array from Figure 9.27.

SUFFIXTREE$_i$(*Text*). For example, this split happens at the purple node while inserting "anamabananas$" into SUFFIXTREE$_3$(*Text*) (Figure 9.28 (top)) and at the purple edge labeled "namabananas$" while inserting "ananas$" into SUFFIXTREE$_4$(*Text*), thus breaking this edge into edges labeled "na" and "mabananas$" in SUFFIXTREE$_5$(*Text*) (Figure 9.28 (bottom)).

STOP and Think: How would you find the node or edge where the partial suffix tree splits while constructing the suffix tree from the suffix array and LCP array?

To find the node/edge where the partial suffix tree splits, we will walk up the rightmost path (i.e., the last added path in the partial suffix tree) in the partial suffix tree beginning at the previously inserted leaf, labeled SUFFIXARRAY(i), to the root. We stop when we encounter the first node v such that DESCENT(v) $\leq$ LCP($i+1$). Afterwards, we have to consider two cases depending on whether DESCENT(v) = LCP($i+1$) (the split occurs at the node v) or DESCENT(v) < LCP($i+1$) (the split occurs on the edge leading from v):

- DESCENT(v) = LCP($i+1$): the concatenation of the labels on the path from the root to v equals the longest common prefix of suffixes corresponding to SUFFIXARRAY(i) and SUFFIXARRAY($i+1$). We insert SUFFIXARRAY($i+1$) as a new leaf x connected to v, and we label the edge (v, x) with the suffix of *Text* starting at position SUFFIXARRAY($i+1$) + LCP($i+1$). Thus, the edge label consists of the remaining symbols of the suffix corresponding to SUFFIXARRAY($i+1$) that are not already represented by the concatenation of the labels of the path connecting the root to v. This completes the construction of the partial suffix tree SUFFIXTREE$_{i+1}$(*Text*) (see Figure 9.28 (top) for an example).

- DESCENT(v) < LCP($i+1$): the concatenation of the labels on the path from the root to v has fewer symbols than the longest common prefix of the suffixes corresponding to SUFFIXARRAY(i) and SUFFIXARRAY($i+1$). The question therefore arises of how to recover these missing symbols. We denote the rightmost edge leading from v in SUFFIXTREE$_i$(*Text*) as (v, w) and argue that the missing symbols represent a prefix of this edge's label. In this case, we split this edge and construct SUFFIXTREE$_{i+1}$(*Text*) as described below (see Figure 9.28 (bottom) for an example):

  1. Delete the edge (v, w) from SUFFIXTREE$_i$(*Text*).
  2. Add a new internal node y and a new edge (v, y) labeled by a substring of *Text* starting at position SUFFIXARRAY($i+1$) + DESCENT(v). The new label is formed by the final LCP($i+1$) − DESCENT(v) symbols of the longest common prefix of SUFFIXARRAY(i) and SUFFIXARRAY($i+1$). Thus, the concatenation of the labels on the path from the root to y is now the longest common prefix of SUFFIXARRAY(i) and SUFFIXARRAY($i+1$).

3. Define DESCENT(y) as LCP($i + 1$).

4. Connect w to the newly created internal node y by an edge (y, w) that is labeled by a substring of *Text* starting at position SUFFIXARRAY(i) + LCP($i + 1$) and ending at position SUFFIXARRAY(i) + DESCENT(w) − 1. The new label consists of the remaining symbols of the deleted edge (v, w) that were not used as the label of edge (v, y).

5. Add SUFFIXARRAY($i + 1$) as a new leaf x as well as an edge (y, x) that is labeled by a suffix of *Text* beginning at position SUFFIXARRAY($i + 1$) + LCP($i + 1$). The label of this edge consists of the remaining symbols of the suffix corresponding to SUFFIXARRAY($i + 1$) that are not already represented by the concatenation of the labels on the path from the root to v.

EXERCISE BREAK: Prove that the running time of this algorithm is $\mathcal{O}(|Text|)$.

Binary search

The game show *The Price is Right* features a timed challenge called the "Clock Game" in which a contestant makes repeated guesses at the price of an item, with the host telling the contestant only whether the true price is higher or lower than the most recent guess.

An intelligent strategy for the Clock Game is to pick a sensible range of prices within which the item's price must fall, and then guess a price halfway between these two extremes. If this guess is incorrect, then the contestant has immediately eliminated half of the set of possible prices. The contestant then makes a guess in the middle of the remaining possible prices, eliminating half of them again. Iterating this strategy quickly yields the price of the item.

This strategy for the Clock Game motivates a binary search algorithm finding the position of an element *key* within a sorted array ARRAY. This algorithm, called **BINARYSEARCH**, is initialized by setting *minIndex* equal to 0 and *maxIndex* equal to the length of ARRAY. It sets *midIndex* equal to (*minIndex* + *maxIndex*)/2 and then checks to see whether *key* is greater than or less than ARRAY(*midIndex*). If *key* is larger than this value, then **BINARYSEARCH** iterates on the subarray of ARRAY from *minIndex* to *midIndex* − 1; otherwise, **BINARYSEARCH** iterates on the subarray of ARRAY from *midIndex* + 1 to *maxIndex*. Iteration eventually identifies the position of *key*.

For example, if *key* = 9 and ARRAY = (1, 3, 7, 8, 9, 12, 15), then **BINARYSEARCH** would first set *minIndex* equal to 0, *maxIndex* equal to 6, and *midIndex* equal to 3. Because *key* is greater than ARRAY(*midIndex*) = 8, we examine the subarray whose elements are greater than ARRAY(*midIndex*) by setting *minIndex* equal to 4, so that *midIndex* is recomputed as $(4+6)/2 = 5$. This time, *key* is smaller than ARRAY(*midIndex*) = 12, and so we examine the subarray whose elements are smaller than this value. This subarray consists of only a single element, which is *key*.

BINARYSEARCH(ARRAY, *key*, *minIndex*, *maxIndex*)
    while *maxIndex* $\geq$ *minIndex*
        midIndex $\leftarrow$ (*minIndex* + *maxIndex*)/2
        if ARRAY(*midIndex*) = *key*
            return *midIndex*
        else if ARRAY(*midIndex*) < *key*
            minIndex $\leftarrow$ *midIndex* + 1
        else
            maxIndex $\leftarrow$ *midIndex* $-$ 1
    return "key not found"

Bibliography Notes

The Aho-Corasick algorithm was introduced by Aho and Corasick, 1975. Suffix trees were introduced by Weiner, 1973. Suffix arrays were introduced by Manber and Myers, 1990. The Burrows-Wheeler Transform was introduced by Burrows and Wheeler, 1994. An efficient implementation of the Burrows-Wheeler transform was described by Ferragina and Manzini, 2000. The genetic cause of Ohdo syndrome was elucidated by Clayton-Smith et al., 2011. Rearrangements and indels in the human genome were studied by Tuzun et al., 2005 and Montgomery et al., 2013. BLAST, the dominant database search tool in molecular biology, was developed by Altschul et al., 1990.

Classifying the HIV Phenotype

How does HIV evade the human immune system?

In 1984, US Health and Human Services Secretary Margaret Heckler declared that an HIV vaccine would be available within two years, stating, "Yet another terrible disease is about to yield to patience, persistence and outright genius."

In 1997, Bill Clinton established a new research center at the National Institutes of Health with the goal of developing an HIV vaccine. In his words, "It is no longer a question of *whether* we can develop an AIDS vaccine, it is simply a question of *when*."

In 2005, Merck began clinical trials of an HIV vaccine but discontinued them two years later after learning that the vaccine actually *increased* the risk of HIV infection in some recipients.

Today, despite enormous investment and ongoing clinical trials, we are still far from an HIV vaccine, and 35 million people are living with the disease. Scientists have made great progress in developing a successful **antiretroviral therapy**, a drug cocktail that stabilizes an infected patient's symptoms. However, this therapy does not cure AIDS and cannot prevent the spread of HIV, and so it does not hold the promise of a true vaccine for containing the AIDS epidemic.

Classical vaccines against viruses are often made from the surface proteins of a virus. These vaccines stimulate the human immune system to recognize viral envelope proteins as foreign, destroy them, and keep a record of it, so that the immune system can identify and eradicate the virus in a later encounter.

However, HIV viral envelope proteins are extremely variable because the virus must mutate rapidly in order to survive (see **DETOUR: The Red Queen Effect**). The HIV population in a *single* infected individual rapidly evolves to evade the human immune system (Figure 10.1), not to mention that HIV strains taken from *different* patients represent multiple highly diverged subtypes. Therefore, a successful HIV vaccine must be broad enough to account for this variability.

In an effort to counteract HIV's variability, we could create a single peptide that contains the least variable segments of the envelope proteins taken from all known HIV strains and use this peptide as the basis for a universal vaccine fighting all HIV strains. However, not only do HIV envelope proteins mutate fast, but they are also "masked" by **glycosylation**, a post-translational modification that often makes these proteins invisible to the human immune system (see **DETOUR: Glycosylation**). As a result, all attempts at developing an HIV vaccine have thus far failed.

HIV has just nine genes, and in this chapter we will focus on the rapidly mutating *env* gene, which has a mutation rate of 1 to 2% per nucleotide per year. The protein

```
VKKLGEQFR-NKTIIFNQPSGGDLEIVMHSFNCGGEFFYCNTTQLFN----------NSTES------DTITL
VKKLGEQFR-NKTIIFNQPSGGDLEIVMHSFNCGGEFFYCNTTQLFN----------NSTDNG-----DTITL
VKKLGEQFR-NKTIIFNQPSGGDLEIVMHSFNCGGEFFYCNTTQLFD----------NSTESNN----DTITL
VDKLREQFGKNKTIIFNQPSGGDLEIVMHTFNCGGEFFYCNTTQLFNSTWNS---TGNGTESYNGQENGTITL
VDKLREQFGKNKTIIFNQPSGGDLEIVMHTFNCGGEFFYCNTTQLFNSTWNG---TNTT--GLDG--NDTITL
VDKLREQFGKNKTIIFNQSSGGDLEIVTHTFNCGGEFFYCNITQLFNSNWIG---NSTE--GLHG--DDTITL
VKKLGEQFG-NKTIIFNQSSGGDLEIVMHSFNCGGEFFYCNTTQLFNN--TR-----NSTESNNGQGNDTTTL
VKKLREQFGKNKTIIFKQSSGGDLEIVTHTFNCAGEFFYCNTTQLFNSNWTE-----NSITGLDG--NDTITL
VGKLREQFGK-KTIIFNQPSGGDLEIVMHSFNCQGEFFYCNTTRLFNSTWDNSTWNSTGKDKENGN-NDTITL
```

FIGURE 10.1 A multiple alignment of a short region of gp120 proteins sampled from a single HIV-positive patient at nine different time points. Almost half of the columns (shown in darker text) are not conserved across all time points, illustrating how quickly HIV evolves, even within an individual host. Amino acids differing from the most common symbol in a column are shown in blue.

encoding the *env* gene then gets cut into **glycoprotein gp120** (approx. 480 amino acids) and **glycoprotein gp41** (approx. 345 amino acids). Together, gp120 and gp41 form the **envelope spike**, which mediates entry of the HIV virus into human cells.

Since HIV mutates so fast, different HIV isolates may have different phenotypes, thus requiring different drug cocktails. For example, HIV viruses can be divided into fast-replicating **syncytium-inducing (SI)** isolates and slow-replicating **non-syncytium-inducing (NSI)** isolates. During infection, viral proteins like gp120 that are used by HIV to enter the cell are transported to the cell surface, where they can cause the host cell membrane to fuse with neighboring cells. This causes dozens of human cells to fuse their cell membranes into a giant, nonfunctional **syncytium**, or abnormal multinucleate cell (Figure 10.2). This mechanism allows an SI virus to kill many human cells by infecting only one.

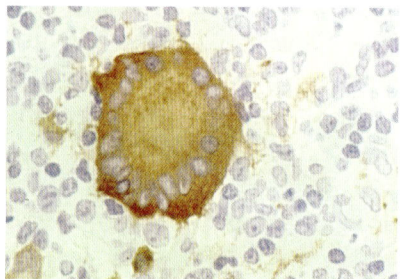

FIGURE 10.2 A syncytium with multiple nuclei in an HIV patient.

Because gp120 is important in classifying a virus as SI or NSI, biologists are interested in determining which amino acids in gp120 can be used for this classification.

In 1992, Jean-Jacques De Jong analyzed a multiple alignment of the **V3 loop** region in gp120 (Figure 10.3 (top)) and devised the **11/25 rule**, which asserts that an HIV strain is more likely to have an SI phenotype if the amino acid at either positions 11 or 25 of its V3 loop is arginine or lysine. It later was shown that many other positions influence the SI/NSI phenotype.

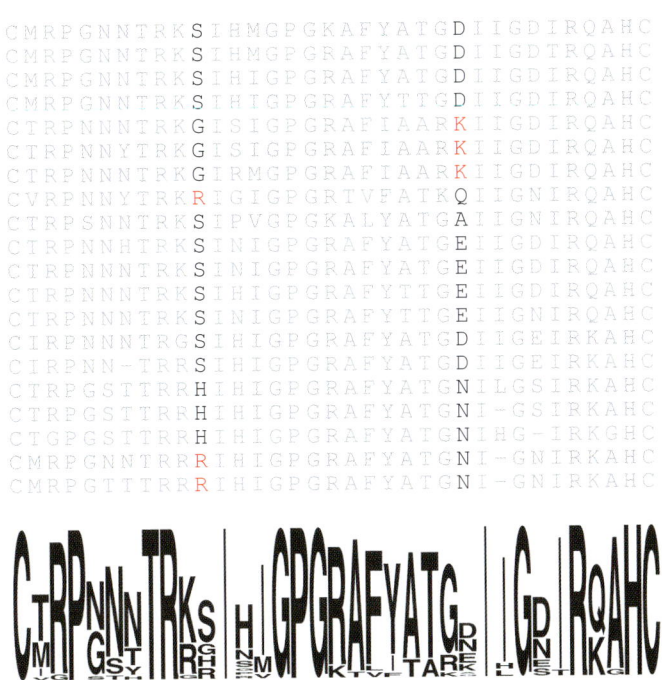

FIGURE 10.3 A multiple alignment of the V3 loop region taken from twenty HIV patients, and the motif logo of this alignment. The alignment's 11th and 25th columns are shown in darker text; occurrences of arginine (R) or lysine (K) in these columns are shown in red. The 11/25 rule will classify six of the patients as infected with an SI isolate. Although the V3 loop is an important and rather conserved segment of gp120, the level of conservation varies across different positions. For example, whereas the first and last positions are extremely conserved, positions 11 and 25 exhibit high variation.

Limitations of sequence alignment

Before biologists could even start to examine the question of predicting HIV phenotypes from gp120 sequences, they faced the problem of constructing accurate multiple alignments of these sequences. Indeed, a single misalignment, placing an incorrect amino acid at a position influencing the SI/NSI phenotype, could lead to a faulty classification of HIV phenotypes. And we already know from Chapter 5 that constructing a multiple alignment of highly diverged sequences is a difficult algorithmic problem.

Figure 10.3 (bottom) shows a motif logo from the V3 loop of gp120 and illustrates that some positions in gp120 are relatively conserved, whereas others are extremely variable. Furthermore, this motif logo does not account for insertions and deletions, which are prevalent in other regions of gp120 that are less conserved than the V3 loop. These insertions and deletions make analyzing gp120 even more complex.

Because the columns of the multiple alignment in Figure 10.3 have varying levels of conservation, we question the wisdom of using the *same* amino acid scoring matrix (as well as indel penalties) across different columns of an alignment. A better approach would use a different scoring approach at *different* columns. For example, an amino acid differing from R in position 3 of the alignment in Figure 10.3 should incur a larger penalty than an amino acid deferring from S in position 11.

In other words, the *problem formulation* of multiple sequence alignment introduced in Chapter 5 does not offer an adequate translation of the biological problem of HIV classification into an algorithmic problem. We must therefore devise a new problem formulation for sequence alignment that will lead to a statistically solid analysis of gp120 proteins. But first, we will ask you to join us in our time machine for one more trip.

Gambling with Yakuza

The Japanese crime syndicates called yakuza descend from groups of 18th Century traveling gamblers called bakuto. (In fact, "yakuza" is the name of a losing hand in a Japanese card game.) One of the most popular games that the bakuto would host in their makeshift casinos is called **Chō-Han**. In this game, which literally translates as "even-odd", the dealer rolls two dice, and players wager on whether the sum of the dice will be even or odd.

Although playing Chō-Han in a yakuza gambling house would undoubtedly make for a fun evening, we can play an equivalent — albeit less exciting — game called "Heads or Tails" by flipping a coin and wagering on the outcome. Assume that for some strange reason, more people wager on tails than on heads in this game. Then a crooked dealer might use a biased coin that is more likely to result in heads than tails. We will assume that this biased coin results in heads with probability 3/4.

STOP and Think: Say that you play Heads or Tails 100 times, and the coin produces heads 63 times. Is the dealer cheating? Was the coin fair or biased?

This question is not well-formulated, since either coin could have produced any sequence of flips. But can we determine which coin is *more likely* to have been used?

We write the probabilities of tails ("T") and heads ("H") for the fair coin (F) as

$$\Pr_F(\text{"H"}) = 1/2 \quad \Pr_F(\text{"T"}) = 1/2$$

and the probabilities for the biased coin (B) as

$$\Pr_B(\text{"H"}) = 3/4 \quad \Pr_B(\text{"T"}) = 1/4$$

Since coin flips are independent events, the probability that n flips of the fair coin will generate a given sequence $x = x_1 x_2 \ldots x_n$ with k occurrences of "H" is

$$\Pr(x|F) = \prod_{i=1}^{n} \Pr_F(x_i) = (1/2)^n.$$

On the other hand, the probability that the biased coin will generate the same sequence is

$$\Pr(x|B) = \prod_{i=1}^{n} \Pr_B(x_i) = (1/4)^{n-k} \cdot (3/4)^k = 3^k/4^n.$$

If $\Pr(x|F) > \Pr(x|B)$, then the dealer more likely used a fair coin, and if $\Pr(x|F) < \Pr(x|B)$, then the dealer more likely used a biased coin. The numbers $(1/2)^n$ and $3^k/4^n$ are so small for large n that in order to compare them, we will use their **log-odds ratio**,

$$\log_2\left(\frac{\Pr(x|F)}{\Pr(x|B)}\right) = \log_2\left(\frac{2^n}{3^k}\right) = n - k \cdot \log_2 3.$$

EXERCISE BREAK: Show that $\Pr(x|F)$ is larger than $\Pr(x|B)$ when the log-odds ratio is positive (i.e., when $k/n < 1/\log_2 3$) and smaller than $\Pr(x|B)$ when the log-odds ratio is negative (i.e., when $k/n > 1/\log_2 3$).

Returning to our example of witnessing $k = 63$ heads in $n = 100$ flips, the log-odds ratio is positive, since

$$k/n = 0.63 < 1/\log_2 3 \approx 0.6309.$$

It follows that $\Pr(x|F) > \Pr(x|B)$, and so the dealer most likely used a fair coin, even though 63 is closer to 75 than it is to 50.

CHAPTER 10

Two Coins up the Dealer's Sleeve

In bakuto gambling houses, a Chō-Han dealer would remove his shirt during play, displaying a tattooed chest, in order to reduce any suspicions of dice tampering. (These tattoos would later become a yakuza tradition.) We will assume, however, that in Heads or Tails, the crooked dealer is wearing a shirt and keeps both coins up a sleeve, secretly changing them back and forth whenever he likes during the sequence of flips. Since he does not want to be caught switching coins, he does so only occasionally.

We will assume that the crooked dealer switches coins with probability 0.1 after each flip. Given a sequence of coin flips, we must determine when the dealer used the biased coin and when he used the fair coin.

Casino Problem:

Given a sequence of coin flips, determine when the crooked dealer used a fair coin and when he used a biased coin.

Input: A sequence $x = x_1 x_2 \ldots x_n$ of coin flips made by two possible coins (F and B).

Output: A sequence $\pi = \pi_1 \pi_2 \ldots \pi_n$, with each π_i being equal to either F or B and indicating that x_i is the result of flipping the fair or biased coin, respectively.

Unfortunately, this problem is poorly stated, since either coin can generate any outcome. Instead, we need to determine the *most likely* sequence of coins used by the dealer.

STOP and Think: Can you reformulate the Casino Problem so that it makes sense?

A well-defined computational problem for finding the most likely sequence of coins used by the dealer should somehow grade different sequences π as better answers than others. One approach to guessing the most likely coin the dealer used for each flip would be to slide a window (of length $t < n$) along the sequence of flips $x = x_1 \ldots x_n$ and then calculate the log-odds ratio under each window. If the log-odds ratio of the window falls below zero, then the dealer most likely used the biased coin inside this window; otherwise, the dealer most likely used the fair coin.

STOP and Think: Do you see any problems with this method?

There are two issues with the window-sliding approach. First, we have no apparent choice for the length of the window. Second, overlapping windows may classify the same outcome as caused by both the fair and biased coins. For example, if x = "HHHHHTTHHHTTTT", then the window $x_1 \ldots x_{10}$ = "HHHHHTTHHH" has a negative log-odds ratio, and the window $x = x_6 \ldots x_{15}$ = "TTHHHTTTTT" has a positive log-odds ratio. So which coin did the dealer use on the flips $x_6 \ldots x_{10}$ = "TTHHH"?

Finding CG-Islands

In the next section, we will improve our method of grading sequences of coin flips. The solution will lead us to a computational paradigm that has been successfully applied to a wide array of bioinformatics problems, including HIV comparison. For now, however, you may still not believe how coin flipping could possibly relate to sequence comparison. Thus, we will briefly describe a different biological problem that more clearly relates to our coin flipping analogy.

In the early 20th Century, Phoebus Levene discovered the four nucleotides making up DNA. At this time, little was known about DNA (Watson & Crick's double helix paper was still half a century away). As a result, Levene doubted that DNA could store genetic information using just a four-letter alphabet, and he hypothesized that DNA comprised nearly equal amounts of adenine, cytosine, guanine, and thymine.

A century later, we know that complementary nucleotides on *opposing* strands of DNA have equal frequencies because of base pairing — ignoring extremely rare base-pairing errors. However, it is not true that nucleotide frequencies are approximately equal on a *single* strand of DNA. For example, different species have widely varying **GC-content**, or the percentage of cytosine and guanine nucleotides in a genome. For example, the human genome's GC-content is approximately 42%.

After accounting for the human genome's skewed GC-content, we might expect that each of the dinucleotides CC, CG, GC, and GG would occur in the human genome with frequency $0.21 \cdot 0.21 = 4.41\%$. However, the frequency of CG in the human genome is only about 1%! This dinucleotide is so rare because of **methylation**, the most common DNA modification, which typically adds a methyl group (CH_3) to the cytosine nucleotide within a CG dinucleotide. The resulting methylated cytosine has the tendency to further deaminate into thymine (see DETOUR: DNA Methylation). As a result of methylation, CG is the least frequent dinucleotide in many genomes.

Nevertheless, methylation is often suppressed around genes in areas called **CG-islands**, where CG appears relatively frequently (Figure 10.4). If you were to sequence a mammalian genome that you knew nothing about, perhaps one of the first things you might do in order to find genes in this genome is look for CG-islands.

STOP and Think: How would you identify CG-islands in a genome?

	A	C	G	T		A	C	G	T
A	0.053	0.079	0.127	0.036	A	0.087	0.058	0.084	0.061
C	0.037	0.058	0.058	0.041	C	0.067	0.063	0.017	0.063
G	0.035	0.075	0.081	0.026	G	0.053	0.053	0.063	0.042
T	0.024	0.105	0.115	0.050	T	0.051	0.070	0.084	0.084

FIGURE 10.4 Dinucleotide frequencies for a collection of CG-islands (left) and non-CG-islands (right) in the human genome computed for a single strand of the X chromosome. Frequencies of CG are shown in red.

A naive approach to search for CG-islands in a genome would slide a window down the genome, declaring windows with higher frequencies of CG as potential CG-islands. The disadvantages of this method are analogous to those of using a sliding window to determine which coin the crooked dealer most likely used at any given point in time. We do not know how long the window should be, and overlapping windows may simultaneously classify the same genomic position as belonging to a CG-island and as not belonging to a CG-island.

Hidden Markov Models

From coin flipping to a Hidden Markov Model

Our goal is to develop a single concept that models both the crooked dealer and the search for CG-islands in a genome. To this end, we will think about the crooked dealer not as a human but as a primitive machine. We do not know how this machine is constructed, but we do know that it proceeds in a sequence of steps; in each step, it is in one of two hidden states, F and B, and it emits a symbol, "H" or "T".

After each step, the machine makes two decisions:

- Which hidden state will I move to next?

- Which symbol will I emit in that state?

The machine answers the first question by choosing randomly among the F and B states, with probability 0.9 of remaining in its current state and probability 0.1 of changing states. The machine answers the second question by choosing between the symbols "H" and "T" with probabilities that depend on the state it is in. In our coin flipping example, the probabilities for state F (0.5 and 0.5) differ from the probabilities for state B (0.75 and 0.25). Our goal is to infer the machine's most likely sequence of states by analyzing the sequence of symbols that it emits.

We have just transformed the dealer into an abstract machine called a **Hidden Markov Model (HMM)**. The only difference between our specialized "coin flipping machine" and the general concept of an HMM is that the latter can have an arbitrary number of states and may have arbitrary probability distributions governing which state to move into and which symbols to emit. In general, an HMM (Σ, *States*, *Transition*, *Emission*) is defined by a set of four objects:

- an alphabet Σ of emitted symbols;

- a set *States* of **hidden states**;

- a $|States| \times |States|$ matrix *Transition* $= (transition_{l,k})$ of **transition probabilities**, where $transition_{l,k}$ represents the probability of moving from state l to state k;

- a $|States| \times |\Sigma|$ matrix *Emission* $= (emission_k(b))$ of **emission probabilities**, where $emission_k(b)$ represents the probability of emitting symbol b from alphabet Σ when the HMM is in state k.

For each state l,

$$\sum_{\text{all states } k} transition_{l,k} = 1$$

and

$$\sum_{\text{all symbols } b \text{ from } \Sigma} emission_l(b) = 1.$$

EXERCISE BREAK: What are Σ, *States*, *Transition*, and *Emission* for the HMM modeling the crooked dealer?

CHAPTER 10

The HMM diagram

As illustrated in Figure 10.5, an HMM can be visualized using an **HMM diagram**, a graph in which every state is represented by a solid node. Solid directed edges connect every pair of nodes, as well as every node to itself. Each such edge is labeled with the transition probability of moving from one state to the other (or remaining in the same state). In addition, the HMM diagram has dashed nodes representing each possible symbol from the alphabet Σ and dashed edges connecting each state to each dashed node. Each such edge is labeled by the probability that the HMM will emit this symbol while in the given state.

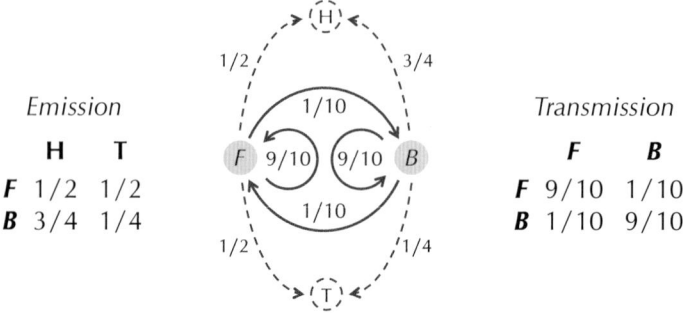

FIGURE 10.5 The transition and emission probability matrices for the crooked dealer HMM described by the HMM diagram shown in the center. This HMM has two states (gray nodes), F and B. In each state, the HMM can emit one of two symbols (dashed nodes), heads ("H") or tails ("T"), with the probabilities shown along dashed edges. Transition probabilities are shown on solid edges; the crooked dealer HMM transitions between states F and B with probability 1/10 and remains in the same state with probability 9/10.

A **hidden path** $\pi = \pi_1 \ldots \pi_n$ in an HMM is the sequence of states that the HMM passes through; such a path corresponds to a path of solid edges in the HMM diagram. Figure 10.6 presents an example in which the crooked dealer HMM produces a sequence of flips x = "THTHHHTHTTH" with hidden path $\pi = FFFBBBBBFFF$, i.e., the fair coin is used for the first three flips and last three flips, and the biased coin is used for the five intermediate flips.

Reformulating the Casino Problem

We can now rephrase the improperly formulated Casino Problem as finding the most likely hidden path π for a string x of symbols emitted by an HMM. To solve this problem,

i	1	2	3	4	5	6	7	8	9	10	11
x	T	H	T	H	H	H	T	H	T	T	H
π	F	F	F	B	B	B	B	B	F	F	F
$\Pr(\pi_i \to \pi_{i+1})$	$\frac{1}{2}$	$\frac{9}{10}$	$\frac{9}{10}$	$\frac{1}{10}$	$\frac{9}{10}$	$\frac{9}{10}$	$\frac{9}{10}$	$\frac{9}{10}$	$\frac{1}{10}$	$\frac{9}{10}$	$\frac{9}{10}$
$\Pr(x_i \mid \pi_i)$	$\frac{1}{2}$	$\frac{1}{2}$	$\frac{1}{2}$	$\frac{3}{4}$	$\frac{3}{4}$	$\frac{3}{4}$	$\frac{1}{4}$	$\frac{3}{4}$	$\frac{1}{2}$	$\frac{1}{2}$	$\frac{1}{2}$

FIGURE 10.6 A sequence x of emitted symbols along with a hidden path π for the crooked dealer HMM. $\Pr(\pi_i \to \pi_{i+1})$ denotes the probability $transition_{\pi_i, \pi_{i+1}}$ of transitioning from state π_i to π_{i+1}. $\Pr(\pi_0 \to \pi_1)$ is set equal to $1/2$ to comply with the assumption that in the beginning, the dealer is equally likely to use the fair or biased coin. $\Pr(x_i|\pi_i)$ denotes the probability that the dealer produced symbol x_i from state π_i and is equal to $emission_{\pi_i}(x_i)$.

we will first consider the simpler problem of computing the probability $\Pr(x, \pi)$ that an HMM follows the hidden path $\pi = \pi_1 \ldots \pi_n$ and emits the string $x = x_1 \ldots x_n$. Note that

$$\sum_{\text{all strings of emitted symbols } x} \sum_{\text{all hidden paths } \pi} \Pr(x, \pi) = 1.$$

STOP and Think: What is $\Pr(x, \pi)$ for the x and π in Figure 10.6?

Each emitted string x has probability $\Pr(x)$, which is independent of the hidden path taken by the HMM:

$$\Pr(x) = \sum_{\text{all hidden paths } \pi} \Pr(x, \pi).$$

Each hidden path π has probability $\Pr(\pi)$, which is independent of the string that the HMM emits:

$$\Pr(\pi) = \sum_{\text{all strings of emitted symbols } x} \Pr(x, \pi).$$

The event "the HMM follows the hidden path π and emits x" can be thought of as a combination of two consecutive events:

- The HMM follows the path π. The probability of this event is $\Pr(\pi)$.

- The HMM emits x, given that the HMM follows the path π. We refer to the probability of this event as the **conditional probability** of x given π, denoted $\Pr(x|\pi)$.

Chapter 10

Both of these events must occur for the HMM to follow path π and emit string x, which implies that

$$\Pr(x, \pi) = \Pr(x|\pi) \cdot \Pr(\pi).$$

PAGE 230 To learn more about this formula, see **DETOUR: Conditional Probability**.

To compute $\Pr(x, \pi)$, we will first compute $\Pr(\pi)$. As shown in Figure 10.6, we write $\Pr(\pi_i \to \pi_{i+1})$ to denote the transition probability of the HMM transitioning from state π_i to π_{i+1}. For simplicity, we assume that in the beginning, the dealer is equally likely to use the fair or biased coin, an assumption that is modeled by setting $\Pr(\pi_0 \to \pi_1) = 1/2$ in Figure 10.6, where π_0 is a "silent" **initial state** that does not emit any symbols. The probability of π is therefore equal to the product of its transition probabilities (purple elements in Figure 10.6),

$$\Pr(\pi) = \prod_{i=1}^{n} \Pr(\pi_{i-1} \to \pi_i) = \prod_{i=1}^{n} transition_{\pi_{i-1}, \pi_i}.$$

Probability of a Hidden Path Problem:

Compute the probability of an HMM's hidden path.

> **Input**: A hidden path π in an HMM (Σ, *States*, *Transition*, *Emission*).
> **Output**: The probability of this path, $\Pr(\pi)$.

Note that we have already computed $\Pr(x|\pi)$ for the crooked dealer HMM when the dealer's hidden path consisted only of B or F, which we wrote as $\Pr(x|B)$ and $\Pr(x|F)$, respectively. To compute $\Pr(x|\pi)$ for a general HMM, we will write $\Pr(x_i|\pi_i)$ to denote the emission probability $emission_{\pi_i}(x_i)$ that symbol x_i was emitted given that the HMM was in state π_i (Figure 10.6). As a result, for a given path π, the HMM emits a string x with probability equal to the product of emission probabilities along that path,

$$\Pr(x|\pi) = \prod_{i=1}^{n} \Pr(x_i|\pi_i)$$
$$= \prod_{i=1}^{n} emission_{\pi_i}(x_i).$$

Probability of an Outcome Given a Hidden Path Problem:
Compute the probability that an HMM will emit a string given its hidden path.

Input: A string $x = x_1 \ldots x_n$ emitted by an HMM (Σ, *States*, *Transition*, *Emission*) and a hidden path $\pi = \pi_1 \ldots \pi_n$.

Output: The conditional probability $\Pr(x|\pi)$ that x will be emitted given that the HMM follows the hidden path π.

Returning to our formula for $\Pr(x, \pi)$, the probability that an HMM follows path π and emits string x can be written as a product of emission and transition probabilities,

$$\Pr(x, \pi) = \Pr(x|\pi) \cdot \Pr(\pi)$$
$$= \prod_{i=1}^{n} \Pr(x_i|\pi_i) \cdot \Pr(\pi_{i-1} \to \pi_i)$$
$$= \prod_{i=1}^{n} emission_{\pi_i}(x_i) \cdot transition_{\pi_{i-1}, \pi_i}.$$

EXERCISE BREAK: Compute $\Pr(x, \pi)$ for the x and π in Figure 10.6. Can you find a better explanation for x = "THTHHHTHTTH" than $\pi = FFFBBBBBFFF$?

STOP and Think: Now that you have learned about HMMs, try designing an HMM that will model searching for CG-islands in a genome. What barriers do you encounter?

The Decoding Problem

The Viterbi graph

As we stated in the previous section, in both the crooked dealer and CG-island HMMs, we are looking for the most likely hidden path π for an HMM that emits a string x. In other words, we would like to maximize $\Pr(x, \pi)$ among all possible hidden paths π.

Decoding Problem:

Find an optimal hidden path in an HMM given a string of its emitted symbols.

Input: A string $x = x_1 \ldots x_n$ emitted by an HMM (Σ, *States*, *Transition*, *Emission*).

Output: A path π that maximizes the probability $\Pr(x, \pi)$ over all possible paths through this HMM.

In 1967, Andrew Viterbi used an HMM-inspired analog of a Manhattan-like grid to solve the Decoding Problem. For an HMM emitting a string of n symbols $x = x_1 \ldots x_n$, the nodes in the HMM's **Viterbi graph** are divided into $|States|$ rows and n columns (Figure 10.7 (middle)). That is, node (k, i) represents state k and the i-th emitted symbol. Each node is connected to all nodes in the column to its right; the edge connecting $(l, i-1)$ to (k, i) corresponds to transitioning from state l to state k (with probability *transition*$_{l,k}$) and then emitting symbol x_i (with probability *emission*$_k(x_i)$). As a result, every path connecting a node in the first column of the Viterbi graph to a node in the final column corresponds to a hidden path $\pi = \pi_1 \ldots \pi_n$.

We assign a weight of

$$\text{WEIGHT}_i(l, k) = transition_{\pi_{i-1}, \pi_i} \cdot emission_{\pi_i}(x_i)$$

to the edge connecting $(l, i-1)$ to (k, i) in the Viterbi graph. Furthermore, we define the **product weight** of a path in the Viterbi graph as the product of its edge weights. For a path from the leftmost column to the rightmost column in the Viterbi graph corresponding to the hidden path π, this product weight is equal to the product of $n - 1$ terms,

$$\prod_{i=2}^{n} transition_{\pi_{i-1}, \pi_i} \cdot emission_{\pi_i}(x_i) = \prod_{i=1}^{n-1} \text{WEIGHT}_i(l, k).$$

STOP and Think: How does this expression differ from the formula for $\Pr(x, \pi)$ that we derived in the previous section?

The only difference between the above expression and the expression that we obtained for $\Pr(x, \pi)$,

$$\prod_{i=1}^{n} transition_{\pi_{i-1}, \pi_i} \cdot emission_{\pi_i}(x_i),$$

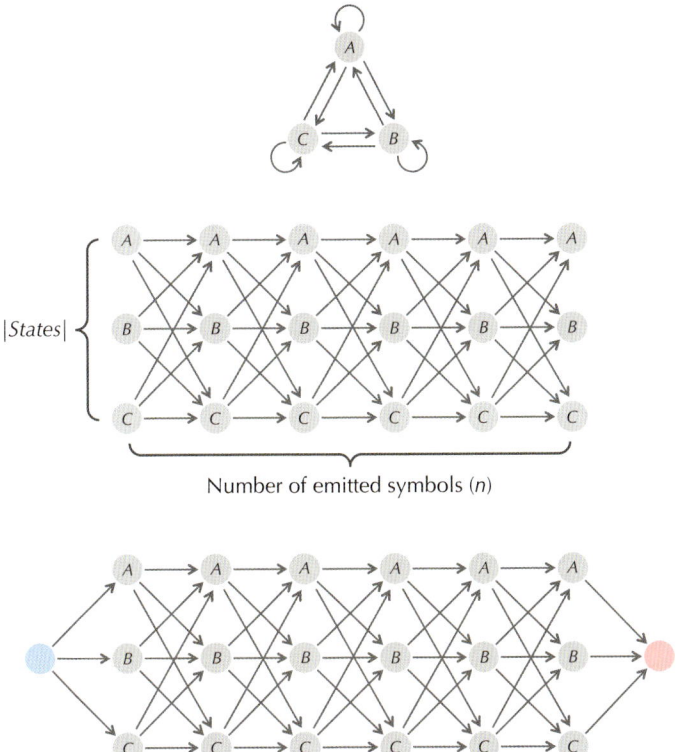

FIGURE 10.7 (Top) The diagram of an HMM with three states (emission/transmission probabilities as well as nodes corresponding to emitted symbols are omitted). (Middle) Given a string of n symbols $x = x_1 \ldots x_n$ emitted by an HMM, Viterbi's Manhattan is a grid with $|States|$ rows and n columns in which each node is connected to every node in the column to its right. The weight of the edge connecting $(l, i-1)$ to (k, i) is $\text{WEIGHT}_i(l, k) = \text{transition}_{l,k} \cdot \text{emission}_k(x_i)$. Unlike the alignment graphs from Chapter 5, in which the set of valid directions was restricted to south, east, and southeast edges, every node in a column is connected by an edge to every node in the column to its right in the Viterbi graph. (Bottom) The Viterbi graph with additional source node (blue) and sink node (red). A path of largest product weight connecting the source to the sink corresponds to an optimal hidden path solving the Decoding Problem.

is the single factor $\text{transition}_{\pi_0, \pi_1} \cdot \text{emission}_{\pi_1}(x_1)$, which corresponds to transitioning from the initial state π_0 to π_1 and emitting the first symbol. To model the initial state, we will add a source node *source* to the Viterbi graph and then connect *source* to each node $(k, 1)$ in the first column with an edge of weight $\text{WEIGHT}_0(source, k) = \text{transition}_{\pi_0, k} \cdot \text{emission}_k(x_1)$. We will also assume that the HMM has another silent

CHAPTER 10

terminal state that the HMM enters when it has finished emitting symbols. To model the terminal state, we add a sink node *sink* to the Viterbi graph and connect every node in the last column to *sink* with an edge of weight 1 (Figure 10.7 (bottom)).

Every hidden path π in the HMM now corresponds to a path from *source* to *sink* in the Viterbi graph with product weight $\Pr(x, \pi)$. Therefore, the Decoding Problem reduces to finding a path in the Viterbi graph of largest product weight over all paths connecting *source* to *sink*.

EXERCISE BREAK: Find the maximum product weight path in the Viterbi graph for the crooked dealer HMM when *x*= "HHTT".

The Viterbi algorithm

We will apply a dynamic programming algorithm to solve the Decoding Problem. First, define $s_{k,i}$ as the product weight of an optimal path (i.e., a path with maximum product weight) from *source* to the node (k, i). The **Viterbi algorithm** relies on the fact that the first $i - 1$ edges of an optimal path from *source* to (k, i) must form an optimal path from *source* to $(l, i - 1)$ for some (unknown) state l. This observation yields the following recurrence:

$$s_{k,i} = \max_{\text{all states } l} \left\{ s_{l,i-1} \cdot (\text{weight of edge between nodes}(l, i-1) \text{ and } (k, i)) \right\}$$

$$= \max_{\text{all states } l} \left\{ s_{l,i-1} \cdot \text{WEIGHT}_i(l, k) \right\}$$

$$= \max_{\text{all states } l} \left\{ s_{l,i-1} \cdot \text{transition}_{\pi_{i-1}, \pi_i} \cdot \text{emission}_{\pi_i}(x_i) \right\}$$

Since *source* is connected to every node in the first column of the Viterbi graph,

$$s_{k,1} = s_{source} \cdot (\text{weight of edge between } source \text{ and } (k, 1))$$

$$= s_{source} \cdot \text{WEIGHT}_0(source, k)$$

$$= s_{source} \cdot \text{transition}_{source, k} \cdot \text{emission}_k(x_1)$$

In order to initialize this recurrence, we set s_{source} equal to 1. We can now compute the maximum product weight over all paths from *source* to *sink* as

$$s_{sink} = \max_{\text{all states } l} s_{l,n}.$$

STOP and Think: How can we adapt our algorithm for finding a longest path in a DAG to find a path with maximum product weight?

How fast is the Viterbi algorithm?

We can interpret the Decoding Problem as yet another instance of the Longest Path in a DAG Problem from Chapter 5 because the path π maximizing the product weight $\prod_{i=1}^{n} \text{WEIGHT}_i(\pi_{i-1}, \pi_i)$ also maximizes the logarithm of this product, which is equal to $\sum_{i=1}^{n} \log\left(\text{WEIGHT}_i(\pi_{i-1})\right)$. Thus, we can substitute the weights of all edges in the Viterbi graph by their logarithms. Finding a longest path (i.e. a path maximizing the *sum* of edge weights) in the resulting graph will correspond to a path of maximum *product* weight in the original Viterbi graph. For this reason, the runtime of the Viterbi algorithm, which you are now ready to implement, is linear in the number of edges in the Viterbi graph. The following exercise shows that the number of these edges is $\mathcal{O}(|States|^2 \cdot n)$, where n is the number of emitted symbols.

EXERCISE BREAK: Show that the number of edges in the Viterbi graph of an HMM emitting a string of length n is $|States|^2 \cdot (n-1) + 2 \cdot |States|$.

EXERCISE BREAK: Apply your solution for the Decoding Problem to find CG-islands in the first million nucleotides from the human X chromosome. To help you design an HMM for this application, you may assume that transitions from CG-islands to non-CG-islands are rare, occurring with probability 0.001, and that transitions from non-CG-islands to CG-islands are even more rare, occurring with probability 0.0001. How many CG-islands do you find?

In practice, many HMMs have **forbidden transitions** between some states. For such transitions, we can safely remove the corresponding edges from the HMM diagram (Figure 10.8 (left)). This operation results in a sparser Viterbi graph (Figure 10.8 (right)), which reduces the runtime of the Viterbi algorithm, since the runtime of the algorithm for finding the longest path in a DAG is linear in the number of edges in the DAG.

EXERCISE BREAK: Let *Edges* denote the set of edges in the diagram of an HMM that may have some forbidden transitions. Prove that the number of edges in the Viterbi graph for this HMM is $|Edges| \cdot (n-1) + 2 \cdot |States|$.

CHAPTER 10

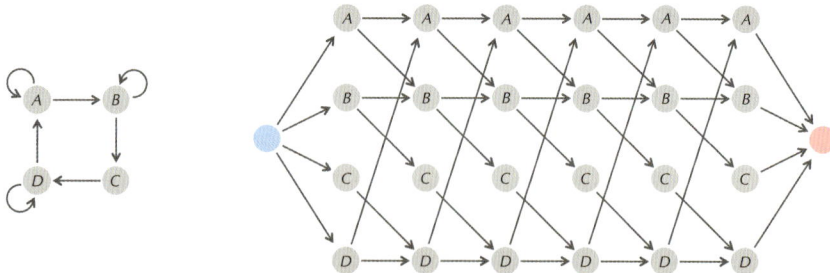

FIGURE 10.8 (Left) An HMM diagram for an HMM that has four states with some forbidden transitions, such as from A to D and from C to itself. Edges corresponding to forbidden transitions between states are not included in the HMM diagram. (Right) The Viterbi graph for this HMM emitting a string of length 6.

Finding the Most Likely Outcome of an HMM

Dynamic programming allows us to answer questions about HMMs extending beyond the most likely hidden path. For example, we have already computed the probability $\Pr(\pi)$ of a hidden path π. But what about computing $\Pr(x)$, the probability that the HMM will emit a string x?

> **EXERCISE BREAK:** Which outcome is more likely in the crooked casino: "HHTT" or "HTHT"? How would you find the most likely sequence of four coin flips?

Outcome Likelihood Problem:
Find the probability that an HMM emits a given string.

Input: A string $x = x_1 \ldots x_n$ emitted by an HMM (Σ, *States*, *Transition*, *Emission*).
Output: The probability $\Pr(x)$ that the HMM emits x.

> **STOP and Think:** To solve the Outcome Likelihood Problem, you can make a slight change to the Viterbi recurrence $s_{k,i} = \max_{\text{all states } l}\{s_{l,i-1} \cdot \text{WEIGHT}_i(l,k)\}$. What is the change?

We have already observed that $\Pr(x)$ is equal to the sum of $\Pr(x, \pi)$ over all hidden paths π. However, the number of paths through the Viterbi graph is exponential in the length of the emitted string x, and so we will use dynamic programming to develop a faster approach to compute $\Pr(x)$.

We denote the total product weight of all paths from *source* to node (k, i) in the Viterbi graph as $forward_{k,i}$; note that $forward_{sink}$ is equal to $\Pr(x)$. To compute $forward_{k,i}$, we will divide all paths connecting *source* to (k, i) into $|States|$ subsets, where each subset contains those paths that pass through node $(l, i-1)$ (with product weight $forward_{l,i-1}$) before reaching (k, i) for some l between 1 and $|States|$. Therefore, $forward_{k,i}$ is the sum of $|States|$ terms,

$$forward_{k,i} = \sum_{\text{all states } l} forward_{l,i-1} \cdot (\text{weight of edge connecting } (l, i-1) \text{ and } (k, i))$$
$$= \sum_{\text{all states } l} forward_{l,i-1} \cdot \text{WEIGHT}_i(l, k).$$

Note that the only difference between this recurrence and the Viterbi recurrence,

$$s_{k,i} = \max_{\text{all states } l} \left\{ s_{l,i-1} \cdot \text{WEIGHT}_i(l, k) \right\},$$

is that the maximization in the Viterbi algorithm has changed into a summation symbol. We can now solve the Outcome Likelihood Problem by computing $forward_{sink}$, which is equal to

$$\sum_{\text{all states } k} forward_{k,n}.$$

Now that we can compute $\Pr(x)$ for an emitted string x, a natural question is to find the most likely such string. In the crooked dealer example, this corresponds to finding the most likely sequence of flips over all possible sequences of fair and biased coins that the dealer could use.

Most Likely Outcome Problem:
Find a most likely string emitted by an HMM.

    Input: An HMM $(\Sigma, States, Transition, Emission)$ and an integer n.
    Output: A most likely string $x = x_1 \ldots x_n$ emitted by this HMM, i.e., a string maximizing the probability $\Pr(x)$ that the HMM will emit x.

CHAPTER 10

EXERCISE BREAK: Solve the Most Likely Outcome Problem (Hint: You may need to build a 3-dimensional version of Viterbi's Manhattan).

Profile HMMs for Sequence Alignment

How do HMMs relate to sequence alignment?

You may still be wondering what in the world HMMs have to do with our original problem of aligning sequences using a column-specific score. As we will see, HMMs offer an elegant solution to this problem.

Given a family of related proteins, we can check whether a newly sequenced protein belongs to this family by constructing pairwise alignments between the newly sequenced protein and each member of the family. If one of the resulting alignments scores above some stringent threshold, then we can assume that the new protein belongs to the family. However, this approach may fail to identify distantly related proteins, such as gp120 proteins taken from different HIV isolates, since these proteins may have scores falling below the threshold. If a sequence has weak similarities with many family members, then it most likely belongs to the family.

The problem, then, is to align a new protein to *all* members of the family at once. To do so, we must assume that we have already constructed a multiple alignment of a family of proteins. Fortunately, it will often be obvious that two proteins come from the same family (e.g., if the proteins are taken from closely related species). Accordingly, biologists often start by constructing an alignment of undeniably related proteins, which are typically easy to align even using the simple multiple alignment methods that we covered in Chapter 5.

Figure 10.9 (first panel) shows a 5×10 alignment *Alignment* representing a hypothetical family of proteins. Note that the sixth and seventh columns of this alignment contain many space symbols and likely do not represent meaningful characteristics of the family. Accordingly, biologists often ignore columns for which the fraction of space symbols is greater than or equal to a **column removal threshold** θ. Column removal results in a 5×8 **seed alignment** (Figure 10.9 (second panel)).

Given a seed alignment *Alignment** representing a family of related proteins, our aim is to build an HMM that realistically models the propensities of symbols in *Alignment** represented by the profile matrix PROFILE(*Alignment**) (Figure 10.9 (third panel)). Rather than thinking about aligning the existing seed alignment to a given string *Text* (representing a new protein), we will instead think about computing the probability

		1	2	3	4	5	6	7	8
Alignment		A A A A A	C F – C D	D D – A D	E A E E E	F – F F F	A C A – – C D – F – – A A A A	D C D – D	F F C C F
*Alignment**		A A A A A	C F – C D	D D – A D	E A E E E	F – F F F	A C F A A	D C D – D	F F C C F
PROFILE(*Alignment**)	A C D E F	1 0 0 0 0	0 2/4 1/4 0 1/4	0 0 3/4 0 0	1/5 0 0 4/5 0	0 0 0 0 1	3/5 1/5 0 0 1/5	0 1/4 3/4 0 0	0 2/5 0 0 3/5

$M_1 \rightarrow M_2 \rightarrow M_3 \rightarrow M_4 \rightarrow M_5 \rightarrow M_6 \rightarrow M_7 \rightarrow M_8$

FIGURE 10.9 A 5×10 multiple alignment *Alignment* (first panel), its 5×8 seed alignment *Alignment** (second panel), the profile matrix PROFILE(*Alignment**) of the seed alignment (third panel), and the diagram of a simple HMM that models this profile (fourth panel). The seed alignment is obtained from the original alignment by ignoring poorly conserved columns (shaded gray); in this case, we ignore columns for which the fraction of space symbols is greater than or equal to the threshold $\theta = 0.35$. To better illustrate the relationship between the alignment and its seed alignment, we have separated the first five columns in the seed alignment from its last three columns and numbered these columns above the original alignment. The match states MATCH(i) are abbreviated as M_i. The HMM only has one possible path; it is initially in state MATCH(1), the transition probability from state MATCH(i) to state MATCH($i+1$) is equal to 1 for all i, and all other transitions are forbidden. Emission probabilities are equal to frequencies in the profile, e.g., emission probabilities for M_2 are 0 for A, 2/4 for C, 1/4 for D, 0 for E, and 1/4 for F.

that the HMM emits *Text*. If the HMM is designed well, then the more similar *Text* is to the strings in *Alignment**, the more likely it will be emitted by the HMM.

We will first construct a simple HMM that treats the columns of *Alignment** as k sequentially linked states called **match states** (Figure 10.9 (fourth panel)), denoted

CHAPTER 10

MATCH(1), ..., MATCH(k). When the HMM enters state MATCH(i), it emits symbol x_i with probability equal to the frequency of this symbol in the i-th column of PROFILE(*Alignment**). The HMM then moves into state MATCH($i+1$) with transition probability equal to 1.

The **similarity score** between *Alignment** and *Text* is the probability Pr(*Text*) that the HMM for *Alignment** emits *Text*. This score is equal to the product of frequencies in PROFILE(*Alignment**) corresponding to each symbol of *Text*. For example, the probability that the HMM in Figure 10.9 emits ADDAFFDF is

$$1 \cdot \frac{1}{4} \cdot \frac{3}{4} \cdot \frac{1}{5} \cdot 1 \cdot \frac{1}{5} \cdot \frac{3}{4} \cdot \frac{3}{5} = 0.003375.$$

STOP and Think: What are the limitations of the HMM in Figure 10.9?

The HMM that we have proposed does score each column in Figure 10.9 differently, and to a degree, the more similar *Text* is to *Alignment**, the higher its similarity score. However, this HMM it is not in keeping with the spirit of HMMs because it has only one hidden path. Furthermore, it offers a simplistic view of multiple alignment because it does not account for insertions and deletions. Finally, it can only "align" *Text* against *Alignment** if the length of *Text* is exactly equal to the number of columns in *Alignment** (Figure 10.10). Yet we will use this limited HMM as the foundation of a more powerful HMM.

	A	C	D	E	F	A	D	F
	A	F	D	A	–	C	C	F
*Alignment**	A	–	–	E	F	F	D	C
	A	C	A	E	F	A	–	C
	A	D	D	E	F	A	D	F
Text	A	D	D	A	F	F	D	F
emission probability	1	1/4	3/4	1/5	1	1/5	3/4	3/5

FIGURE 10.10 Aligning *Text* = ADDAFFDF against the seed alignment *Alignment** represented as a simple HMM in Figure 10.9. This HMM is limited because we are not able to align a string of length other than 8. Indeed, there is no way to add space symbols to *Text* or to add symbols of *Text* "between" columns of *Alignment**.

Building a profile HMM

The improved HMM that we propose is called a **profile HMM**. Given a multiple alignment *Alignment* and a column removal threshold θ used to obtain a seed alignment *Alignment**, we will denote this profile HMM as HMM(*Alignment*, θ). Because the profile HMM will be constructed from the seed alignment, we will also informally refer to it as HMM(*Alignment**). Given a string *Text* to align against the existing seed alignment, our goal is to find an optimal hidden path in the profile HMM by solving the Decoding Problem for this HMM and the emitted string *Text*.

As with our first attempt at an HMM from Figure 10.9, the profile HMM will still traverse its states in an order consistent with traversing the columns of *Alignment** from left to right. However, to align strings *Text* of varying lengths, we will need more states in addition to the k match states.

First, we add $k + 1$ **insertion states**, denoted INSERTION(0), ..., INSERTION(k) (Figure 10.11). Entering INSERTION(i) allows the profile HMM to emit an additional symbol after visiting the i-th column of PROFILE(*Alignment**) and before entering the $(i + 1)$-th column. Thus, we will connect MATCH(i) to INSERTION(i) and INSERTION(i) to MATCH($i + 1$). Furthermore, to allow for multiple inserted symbols between columns of PROFILE(*Alignment**), we will connect INSERTION(i) to itself.

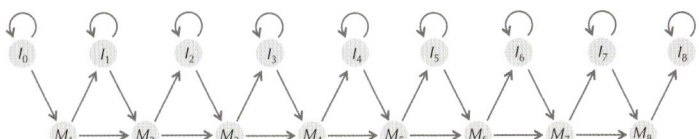

FIGURE 10.11 An HMM diagram for the seed alignment in Figure 10.9 with match and insertion states, abbreviated as *M* and *I*, respectively. The states I_0 and I_8 model insertions of symbols occurring before the beginning and end of *Alignment**, respectively.

STOP and Think: Can we use the HMM in Figure 10.11 to align a string *Text* of length less than 8?

After modeling insertions of new symbols in PROFILE(*Alignment**), we should also model "deletions" allowing the profile HMM to skip columns of PROFILE(*Alignment**). One way of modeling these deletions is to add edges connecting every state in the profile HMM to every state on its right (Figure 10.12).

CHAPTER 10

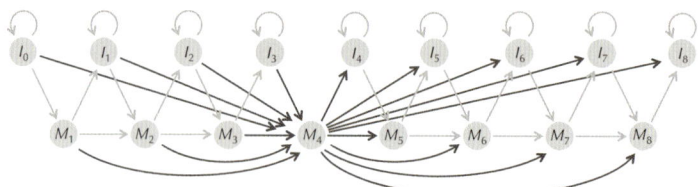

FIGURE 10.12 By adding edges connecting each state in the profile HMM from Figure 10.11 to every state on its right, we can skip columns of *Alignment* when comparing *Text* against this alignment. The above HMM diagram highlights all edges leading into and out of MATCH(4).

STOP and Think: Revisit the Exercise Break on page 195 to recall that the running time of the Viterbi algorithm is proportional to the number of edges (with non-zero transition probabilities) in the HMM diagram. How many edges will the diagram in Figure 10.12 have? How can we reduce the number of edges in the HMM diagram?

Instead of skipping states as in Figure 10.12, we can reduce the number of edges in the HMM diagram by introducing k silent **deletion states** DELETION(1), ..., DELETION(k) (Figure 10.13). For example, instead of jumping from MATCH($i-1$) to MATCH($i+1$), we can make the transition MATCH($i-1$) → DELETION(i) → MATCH($i+1$). Entering DELETION(i) allows the HMM to skip over a column of the alignment without emitting a symbol.

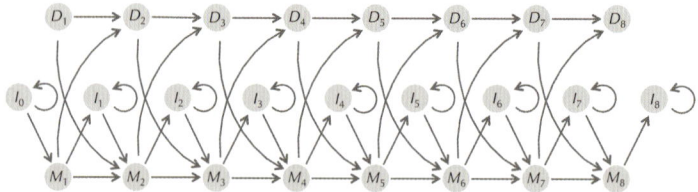

FIGURE 10.13 Adding silent deletion states (abbreviated as D_i) to the profile HMM diagram.

STOP and Think: Is the HMM in Figure 10.13 now adequate, or is there anything else that we have forgotten to add?

We can now transition back and forth between match states and insertion states, as well as back and forth between match states and deletion states, but we cannot transition between insertion states and deletion states. The profile HMM diagram should therefore include edges connecting INSERTION(i) to DELETION($i + 1$) and connecting DELETION(i) to INSERTION(i) for each i. As a result, the profile HMM can move from any match/insertion state to any other match/insertion state on its right by sidetracking through intermediate deletion states. We obtain the complete profile HMM diagram shown in Figure 10.14 after connecting the initial state (S) to the first match/insertion/deletion states and connecting the final match/insertion/deletion states to the terminal state (E).

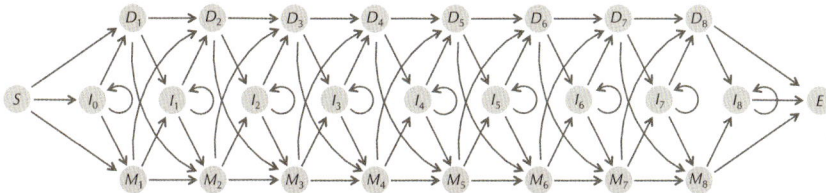

FIGURE 10.14 Adding transitions from insertion states to deletion states and vice-versa completes the profile HMM diagram for the profile matrix from Figure 10.9. Silent initial and terminal states are shown by S and E, respectively.

STOP and Think: Consider the following questions.

- How many edges does the HMM diagram in Figure 10.14 have? How does this compare to the HMM diagram in Figure 10.12?

- What does the Viterbi graph of the profile HMM in Figure 10.14 look like? How many nodes and edges does it have?

Transition and emission probabilities of a profile HMM

In Figure 10.15, we return to the multiple alignment *Alignment* from Figure 10.9 and represent each of the five colored rows of this alignment as a path in the diagram of HMM(*Alignment**). Symbols in the seed alignment *Alignment** (non-shaded columns) correspond to either a match state (non-space symbols) or a deletion state (space symbols). As for symbols not present in the seed alignment (shaded columns), space symbols are ignored, and non-space symbols are emitted from insertion states.

CHAPTER 10

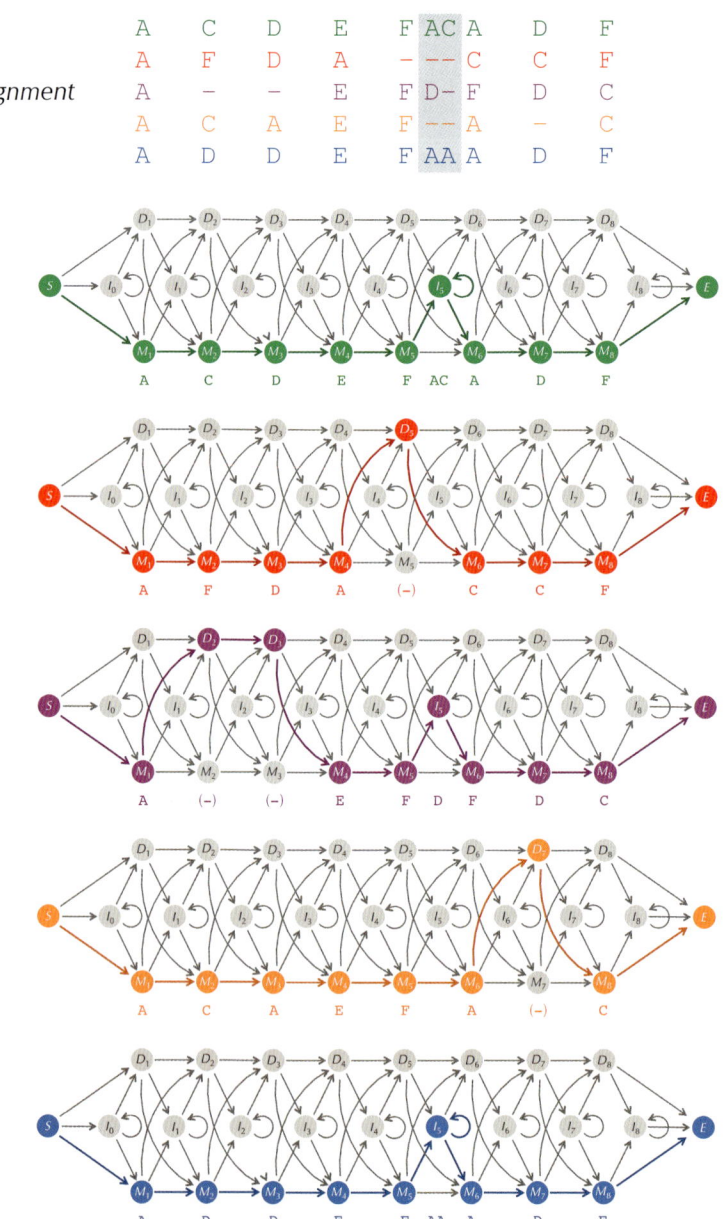

FIGURE 10.15 Five paths through the profile HMM corresponding to the five rows in the alignment from Figure 10.9. Space symbols below an HMM diagram correspond to deletion states and are shown in parentheses to indicate that they are not emitted by the HMM.

STOP and Think: How would you assign transition and emission probabilities for the profile HMM of the alignment in Figure 10.15?

To assign the transition probability $transition_{l,k}$, we simply take the frequency of transitions from state l to state k made by these colored paths with respect to all paths that visited state l. For example, in Figure 10.15, four of the colored paths visit MATCH(5). Three of these paths then transition to INSERTION(5), and one transitions to MATCH(6). Thus, we set the following transition probabilities leaving MATCH(5):

$$transition_{\text{MATCH}(5),\text{INSERTION}(5)} = 3/4$$
$$transition_{\text{MATCH}(5),\text{MATCH}(6)} = 1/4$$
$$transition_{\text{MATCH}(5),\text{DELETION}(6)} = 0$$

We can define the transition probabilities from the initial state analogously. For the multiple alignment in Figure 10.15, we enter MATCH(1) with probability 1; for a general profile HMM, the only other states we could enter from the initial state are INSERTION(0) and DELETION(1). The complete matrix of transmission probabilities is shown in Figure 10.16.

STOP and Think: Due to the small number of strings in the alignment from Figure 10.15, many of the transition probabilities in the gray cells in Figure 10.16 are equal to zero. What are the possible negative consequences of these zeroes, and how would you address these consequences?

To assign the emission probability $emission_k(b)$, we divide the number of times that symbol b was emitted from state k by the total number of symbols emitted from state k. For example, in Figure 10.15, there are three occurrences of A, one occurrence of C, and one occurrence of D emitted from the state INSERTION(5). Also, there are two occurrences of C, one occurrence of D, and one occurrence of F emitted from MATCH(2). We can therefore infer the following emission probabilities for these two states:

$$emission_{\text{INSERTION}(5)}(\text{A}) = 3/5 \qquad emission_{\text{MATCH}(2)}(\text{A}) = 0$$
$$emission_{\text{INSERTION}(5)}(\text{C}) = 1/5 \qquad emission_{\text{MATCH}(2)}(\text{C}) = 2/4$$
$$emission_{\text{INSERTION}(5)}(\text{D}) = 1/5 \qquad emission_{\text{MATCH}(2)}(\text{D}) = 1/4$$
$$emission_{\text{INSERTION}(5)}(\text{E}) = 0 \qquad emission_{\text{MATCH}(2)}(\text{E}) = 0$$
$$emission_{\text{INSERTION}(5)}(\text{F}) = 0 \qquad emission_{\text{MATCH}(2)}(\text{F}) = 1/4$$

CHAPTER 10

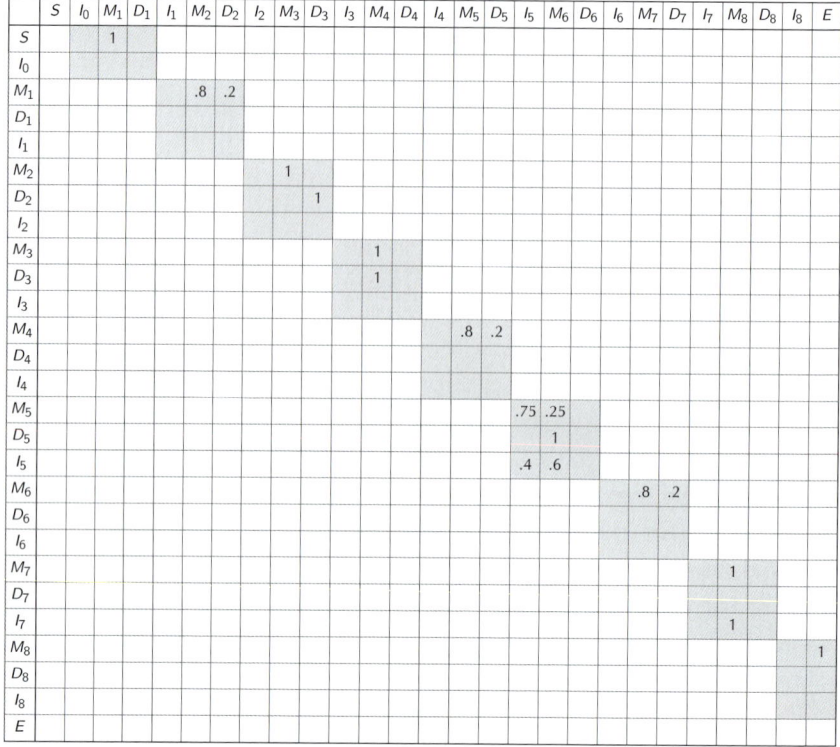

FIGURE 10.16 The 27 × 27 matrix of transition probabilities for HMM(*Alignment*, 0.35), where *Alignment* is the multiple alignment from Figure 10.9. All values in empty cells are equal to zero. Cells shaded gray correspond to edges in the HMM diagram from Figure 10.14; cells shaded white correspond to forbidden transitions.

EXERCISE BREAK: Construct the 27 × 20 emission probability matrix for HMM(*Alignment*, 0.35) derived from *Alignment* in Figure 10.9.

You are now ready to construct the profile HMM for an arbitrary multiple alignment.

Profile HMM Problem:

Construct a profile HMM from a multiple alignment.

Input: A multiple alignment *Alignment* and a threshold θ.
Output: HMM(*Alignment*, θ).

WHY HAVE BIOLOGISTS STILL NOT DEVELOPED AN HIV VACCINE?

EXERCISE BREAK: Construct a profile HMM for the HIV sequences shown in Figure 10.1 with $\theta = 0.35$.

Classifying proteins with profile HMMs

Aligning a protein against a profile HMM

Given a protein family, represented by *Alignment*, we can now return to the problem of deciding whether a newly sequenced protein, represented by *Text*, belongs to the family. We first form HMM(*Alignment*, θ) for some parameter θ. As shown in Figure 10.17, a hidden path through HMM(*Alignment*, θ) corresponds to a sequence of match, insertion, and deletion states for aligning *Text* against *Alignment*.

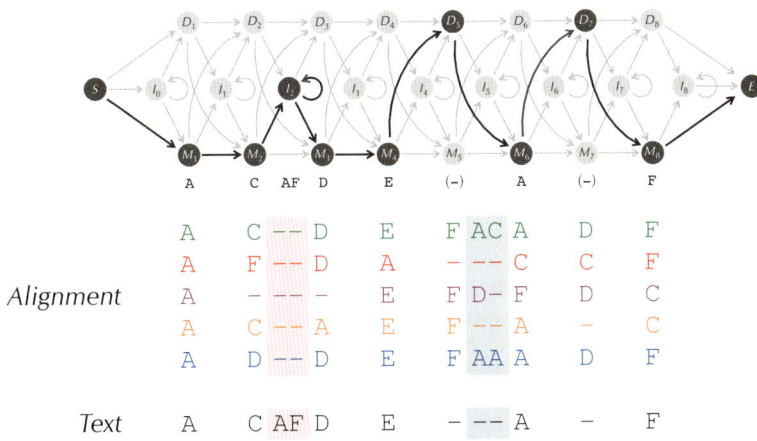

FIGURE 10.17 (Top) A path through HMM(*Alignment*, 0.35) for the multiple alignment from Figure 10.9 and the emitted string *Text* = ACAFDEAF. (Bottom) The emitted symbols correspond to aligning *Text* against *Alignment*. Specifically, the first two symbols are emitted from two match states and belong in the first two positions of the alignment. The next two symbols are emitted from an insertion state and belong in columns of their own (shown in pink). The space symbols in the seventh and eleventh columns above correspond to deletion states; these symbols are not emitted by the HMM. The space symbols in the gray columns do not correspond to any states and are passed over. The non-shaded columns form an augmented 6 × 8 seed alignment for comparison against newly sequenced proteins.

CHAPTER 10

To find the "best" alignment of *Text* against *Alignment*, we simply need to apply the Viterbi algorithm to find an optimal hidden path in HMM(*Alignment*, θ). If the product weight of this optimal hidden path exceeds a predetermined threshold, then we may conclude that *Text* belongs to the protein family, in which case we augment the existing seed alignment with an additional row corresponding to *Text*. In this way, we can recruit more and more distant family members to a seed alignment, adding these new proteins to the growing multiple alignment, and thus making the resulting profile HMM more and more suitable for analyzing the protein family of interest.

STOP and Think: If the product weight for a new protein exceeds a threshold for more than one protein family, how would you classify this protein?

Profile HMMs have finally helped us achieve our original goal of scoring different columns of a multiple alignment differently based on the frequency of symbols in each column. For example, say that the seventh column of *Alignment** contains more occurrences of A than C, and the ninth column of *Alignment** contains more occurrences of C than A. A hidden path passing through MATCH(7) would be rewarded more for emitting A than C, whereas a hidden path passing through MATCH(9) would be rewarded more for emitting C than A.

The return of pseudocounts

The majority of transition probabilities in the gray cells of Figure 10.16 are equal to zero. (The same is true of emission probabilities.) These zeroes may cause problems; for example, the path in Figure 10.17 seems perfectly reasonable for *Text* = ACAFDEAF, and yet Pr(x, π) is equal to zero because the transition probability from MATCH(2) to INSERTION(2) for this profile HMM is zero.

As in Chapter 2, we will introduce pseudocounts by adding a small value σ to entries in the transition matrix that correspond to edges of the HMM diagram in Figure 10.14 (i.e., only the gray elements of Figure 10.16). Note that white cells in Figure 10.16, corresponding to forbidden transitions, are not affected by pseudocounts. The resulting matrix will then need to be normalized so that the elements in each row sum to 1.

EXERCISE BREAK: Compute the normalized matrix for the matrix in Figure 10.16 after adding the pseudocount $\sigma = 0.01$.

We will also add pseudocounts to the matrix of emission probabilities and normalize the resulting matrix. We refer to the profile HMM defined by the resulting normalized matrices of transition and emission probabilities as HMM(*Alignment*, θ, σ).

Profile HMM with Pseudocounts Problem:
Construct a profile HMM with pseudocounts from a multiple alignment.

Input: A multiple alignment *Alignment*, a threshold value θ, and a pseudocount value σ.
Output: HMM(*Alignment*, θ, σ).

STOP and Think: Since the HMM diagram in Figure 10.17 has 25 nodes — not including the start and end states — the Viterbi graph for the string emitted in this figure has 25 rows. How many columns does this Viterbi graph have?

We are now ready to align a string *Text* to a multiple alignment by constructing the Viterbi graph for this string (Figure 10.18) and solving the Decoding Problem to find the most likely hidden path.

STOP and Think: Find paths through the Viterbi graph corresponding to the bottom four hidden paths in Figure 10.15. What happens?

The troublesome silent states

If you reached this point without any questions about Figure 10.18, then we have successfully concealed from you that solving the Decoding Problem for HMMs with silent states is not as simple as it may appear: the graph in Figure 10.18 is not a Viterbi graph! To see why not, consider the path in Figure 10.19, which emits the same string as Figure 10.18 but passes through one fewer silent deletion state, thus reducing the number of columns by one. But the Viterbi graph is not allowed to change depending on the hidden path π, since we know nothing about the hidden path in advance! Instead, the number of columns in the Viterbi graph must equal the length of the *emitted string*, a condition that is violated in both Figure 10.18 and Figure 10.19.

STOP and Think: How can we modify the notion of the Viterbi graph for HMMs with silent states?

More generally, the Viterbi algorithm does not tolerate silent states other than the initial and terminal states. In other words, this algorithm assumes that node (k, i) in the

CHAPTER 10

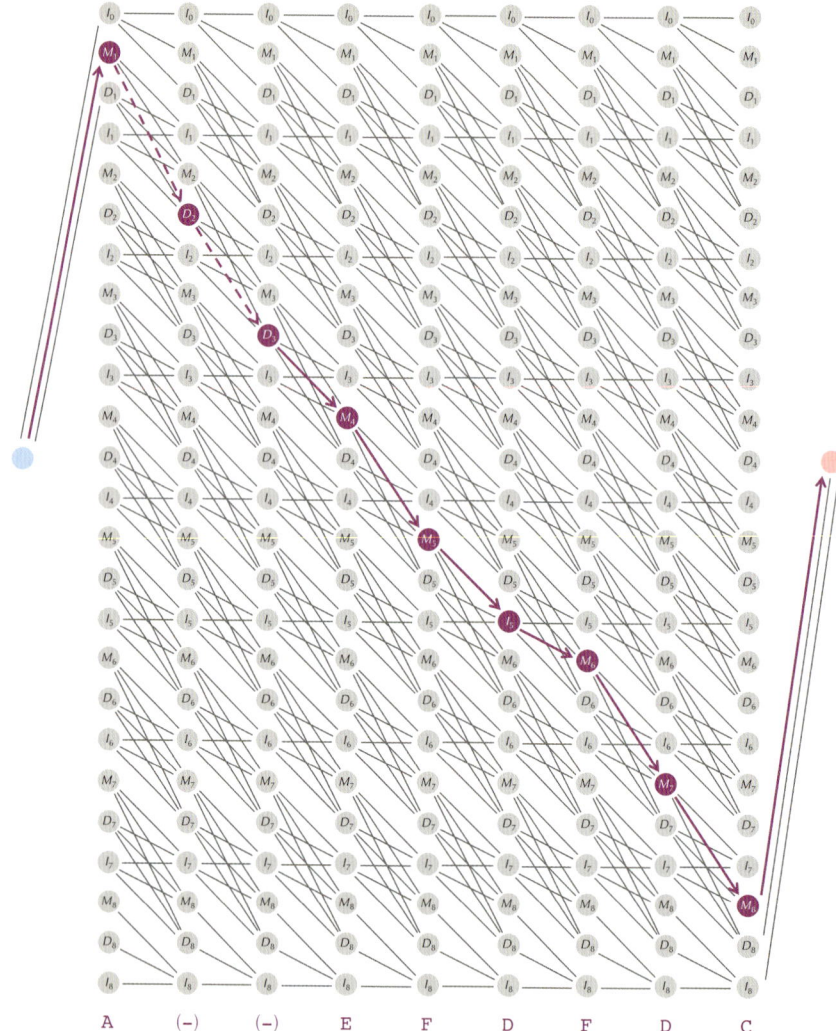

FIGURE 10.18 The Viterbi graph for HMM(*Alignment*, θ) and a path in this graph (shown in purple) corresponding to the hidden path for the emitted string **AEFDFDC** from Figure 10.15. Edges between columns correspond to allowed transitions in the HMM diagram from Figure 10.14 and have an implied rightward orientation. Edges entering nodes corresponding to deletion states are dashed. Emitted symbols are shown beneath each column.

WHY HAVE BIOLOGISTS STILL NOT DEVELOPED AN HIV VACCINE?

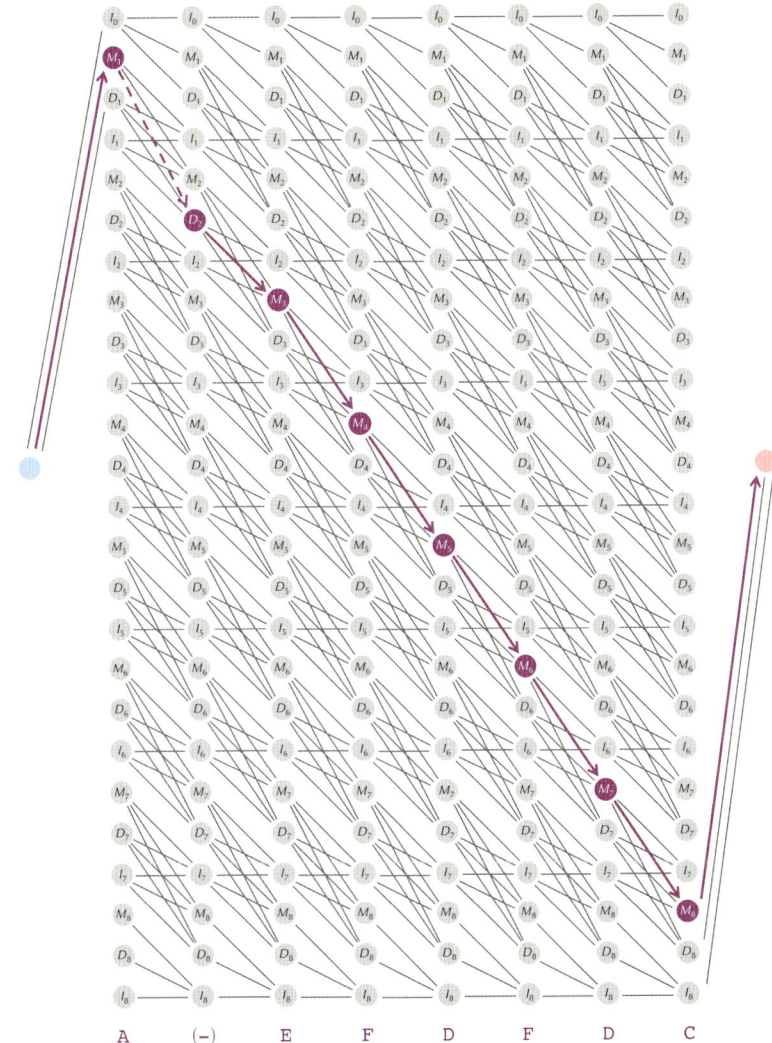

FIGURE 10.19 Another path through another "Viterbi graph" emitting the same string **AEFDFDC** as in Figure 10.18.

CHAPTER 10

Viterbi graph describes the event "the HMM emitted symbol x_i when it was in state k". However, if k is a silent state, then the role of the node (k, i) in the Viterbi graph is poorly defined, as it is unclear how to define the weight of edges entering this node.

Fortunately, we can fix this issue in the case of profile HMMs by defining the Viterbi graph with $|States|$ rows and $|Text|$ columns (Figure 10.20). Every time the HMM moves into a deletion state, rather than crossing over to the next column of the Viterbi graph (as in Figure 10.18 and Figure 10.19), we will move *within* the same column. When the HMM moves into a match or insertion state, we will move to the next column. As a result, every column of the Viterbi graph corresponds to a single emitted symbol, even though a path can pass through more than one state in a given column.

EXERCISE BREAK: Show that the vertical edge connecting (i, l) to (i, k), where k is a deletion state, should be assigned weight equal to $transition_{l,k}$.

STOP and Think: Are there any remaining issues with the graph in Figure 10.20?

There is still a minor flaw with the graph in Figure 10.20. If the HMM moves from the initial state into DELETION(1), then the HMM will move through the first column without emitting a symbol. We will therefore transform the initial state into a column of silent states containing the initial state and all deletion states (Figure 10.21). This way, if the HMM enters DELETION(1) from the initial state, it can move downward through deletion states before transitioning to a match or insertion state in the first column.

You are now ready to use the profile HMM to align a sequence against a seed alignment. The only remaining snare is that when computing $s_{k,i}$ — or, equivalently, $\log(s_{k,i})$ — we must make sure that all incoming scores have been computed. We therefore suggest the top-down, column-by-column topological ordering for the profile HMM shown in Figure 10.22.

Sequence Alignment with Profile HMM Problem:

Align a new sequence to a family of sequences using a profile HMM.

Input: A multiple alignment *Alignment*, a threshold θ, a pseudocount value σ, and a string *Text*.
Output: An optimal hidden path emitting *Text* in HMM(*Alignment*, θ, σ).

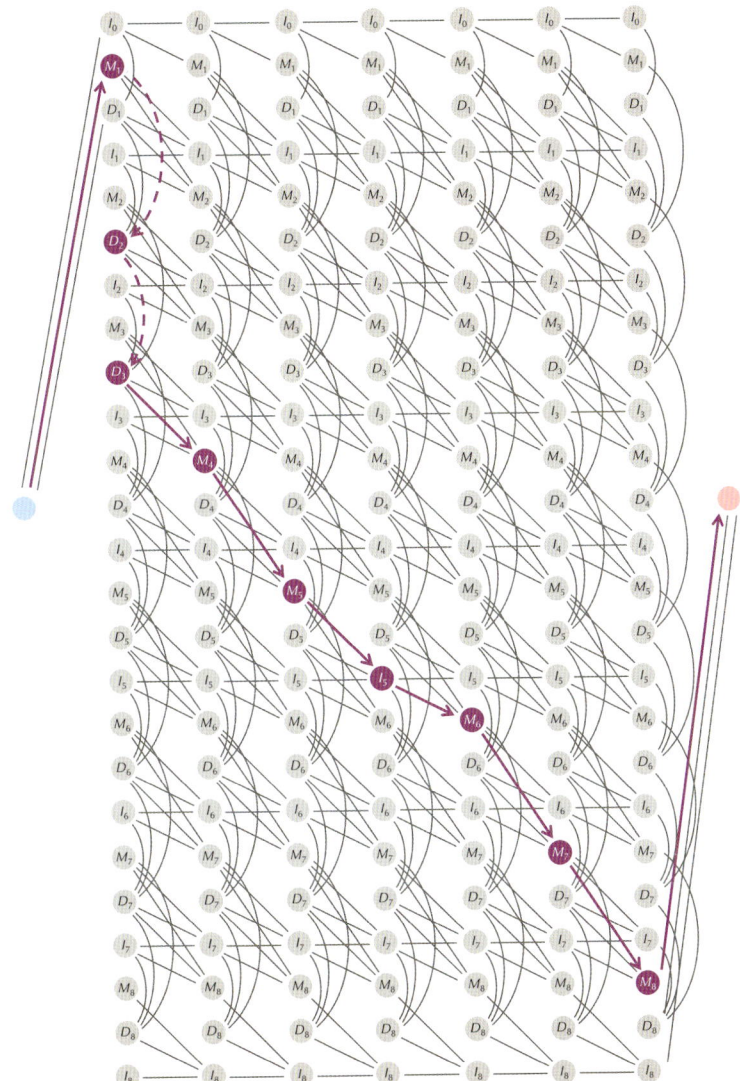

FIGURE 10.20 The Viterbi graph with |*States*| rows and |*Text*| columns for the profile HMM from Figure 10.14 emitting a string *Text* of length 7 so that edges entering deletion states are drawn downward within the same column instead of between columns as in Figure 10.18 and Figure 10.19. The purple path corresponds to the path through the HMM in Figure 10.15 emitting **AEFDFDC**.

CHAPTER 10

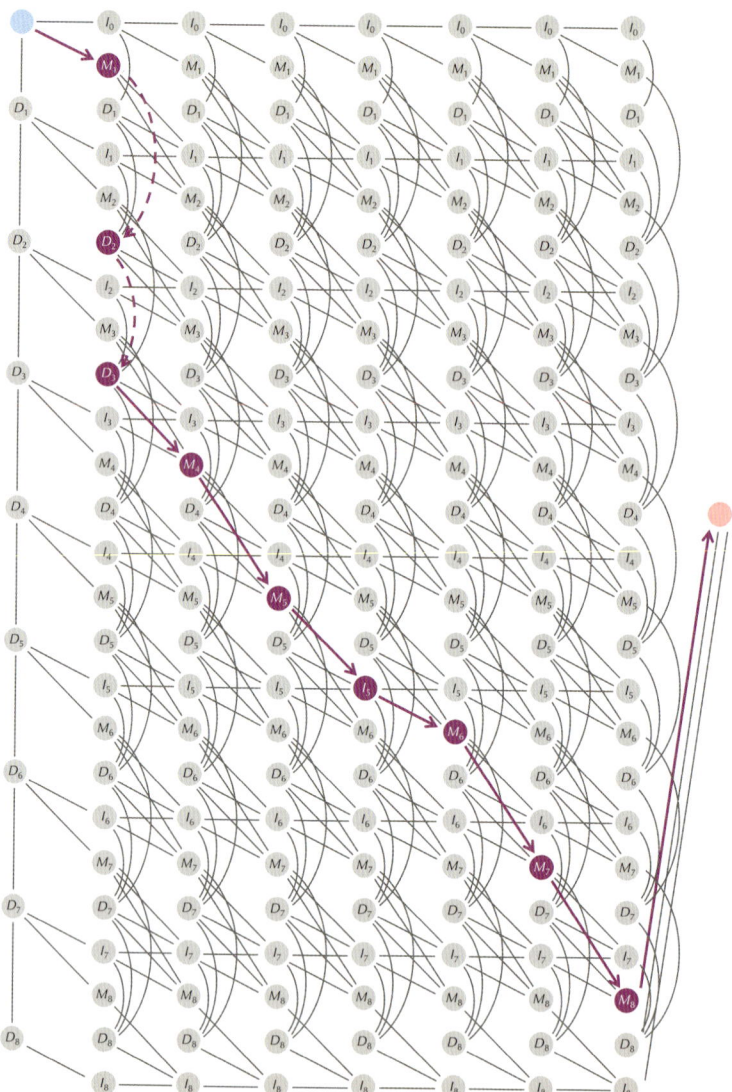

FIGURE 10.21 The final Viterbi graph for the profile HMM in Figure 10.14 emitting a string of length 7. Edges within the same column have a downward orientation; edges between columns have a rightward orientation. Once again, the purple path corresponds to the path through the HMM in Figure 10.15 emitting **AEFDFDC**.

WHY HAVE BIOLOGISTS STILL NOT DEVELOPED AN HIV VACCINE?

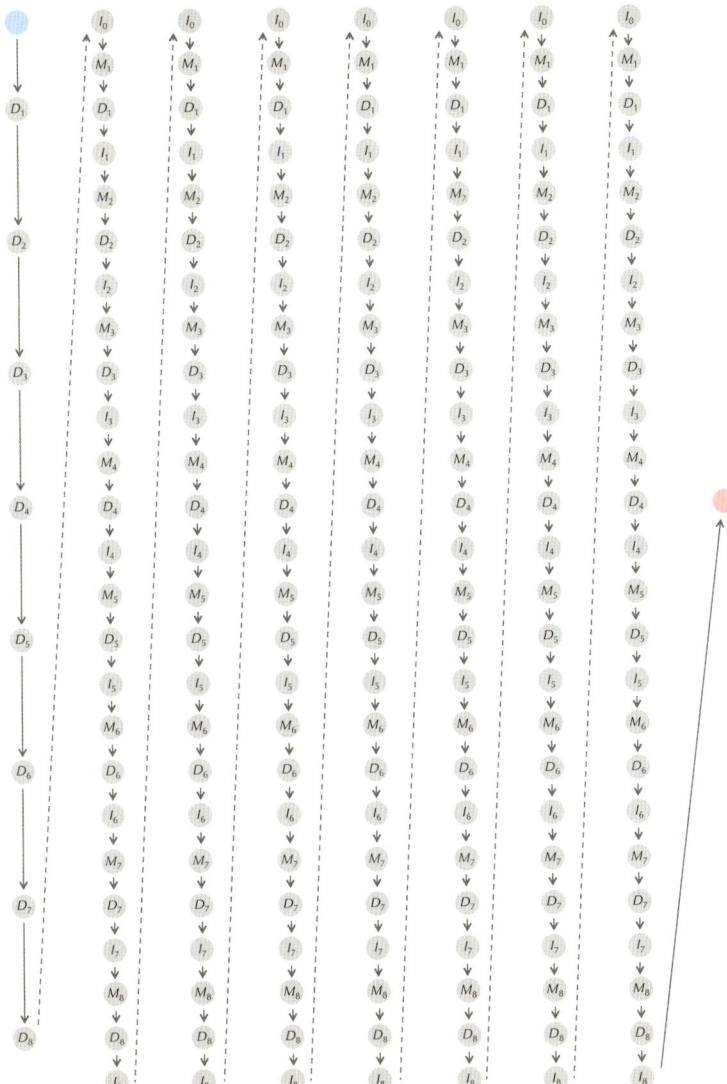

FIGURE 10.22 The topological order of the Viterbi graph from Figure 10.21, which proceeds top-down and column-by-column.

CHAPTER 10

EXERCISE BREAK: Solve the Sequence Alignment with Profile HMM Problem for the profile HMM you constructed for the multiple alignment in Figure 10.1 along with a gp120 protein taken from chimpanzee Simian Immunodeficiency Virus (SIV).

STOP and Think: How would you build the Viterbi graph for an arbitrary HMM with silent states? In which situations will it be impossible to construct a Viterbi graph of an HMM with silent states?

Are profile HMMs really all that useful?

The Viterbi algorithm applies for any HMM, but we will describe how it works for profile HMMs in order to make an important point. Define $s_{\text{MATCH}(j),i}$ as the probability of the most likely hidden path for the prefix $x_1 \ldots x_i$ of x that ends at state $\text{MATCH}(j)$, and define $s_{\text{INSERTION}(j),i}$ and $s_{\text{DELETION}(j),i}$ analogously. Because there are only three edges entering $\text{MATCH}(j)$, the Viterbi recurrence states that

$$s_{\text{MATCH}(j),i} = \max \begin{cases} s_{\text{MATCH}(j-1),i-1} \cdot \text{WEIGHT}_i(\text{MATCH}(j-1), \text{MATCH}(j)) \\ s_{\text{INSERTION}(j-1),i-1} \cdot \text{WEIGHT}_i(\text{INSERTION}(j-1), \text{INSERTION}(j)) \\ s_{\text{DELETION}(j-1),i-1} \cdot \text{WEIGHT}_i(\text{DELETION}(j-1), \text{DELETION}(j)) \end{cases}$$

After taking the logarithm of both sides, the resulting recurrence is very similar to the standard recurrence relation for global pairwise alignment because it is the maximum of three sums:

$$\log\left(s_{\text{MATCH}(j),i}\right) =$$

$$\max \begin{cases} \log\left(s_{\text{MATCH}(j-1),i-1}\right) + \log\left(\text{WEIGHT}_i(\text{MATCH}(j-1), \text{MATCH}(j))\right) \\ \log\left(s_{\text{INSERTION}(j-1),i-1}\right) + \log\left(\text{WEIGHT}_i(\text{INSERTION}(j-1), \text{INSERTION}(j))\right) \\ \log\left(s_{\text{DELETION}(j-1),i-1}\right) + \log\left(\text{WEIGHT}_i(\text{DELETION}(j-1), \text{DELETION}(j))\right) \end{cases}$$

Figure 10.23 shows how a path in a Manhattan-like alignment graph corresponds to a path through the profile HMM. Diagonal edges, vertical edges, and horizontal edges in the Manhattan-like graph correspond to match states, insertion states, and deletion states, respectively.

Figure 10.23 may make it seem that we have wasted your time introducing HMMs, since it appears that a profile HMM is somehow equivalent to pairwise sequence alignment. However, keep in mind that the choice of edges in Figure 10.23 is based on varying transition and emission probabilities. By deriving individual scoring parameters for each column in the alignment matrix, profile HMMs allow us to capture subtle similarities that can fly under the radar of the simple scoring approaches from Chapter 5.

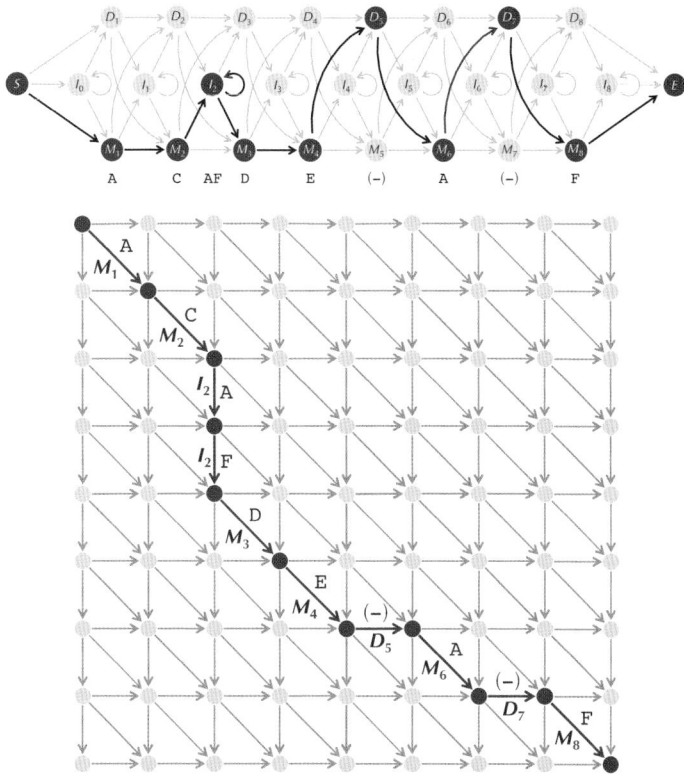

FIGURE 10.23 (Top) The hidden path through the profile HMM in Figure 10.17 (top) emitting ACAFDEAF. (Bottom) The path through a Manhattan-like graph corresponding to this hidden path.

Learning the Parameters of an HMM

Estimating HMM parameters when the hidden path is known

Thus far, our analysis has assumed that we know the parameters of an HMM, i.e., its transition and emission probabilities. We have described a naive — and not necessarily optimal — heuristic for choosing these parameters for a profile HMM, but it is not clear how to select parameters for an arbitrary HMM.

Indeed, the largest complication when modeling biological problems with an HMM is estimating the HMM's parameters from data. In terms of the crooked casino, imagine that you know that the dealer is using two coins to cheat, but you don't know the bias of the coins or the probability that the dealer switches coins at any given time. Can you infer these parameters just from the sequence of coin flips?

STOP and Think: Say that you observe the sequence of coin flips "HHTHH-HTHHTTTH". What is your best guess for the biases of the two coins and for the probabilities of switching from one coin to the other? Would your best guess change if you knew that the hidden path were $\pi = FFFBBFFFFFBBB$?

We will collectively refer to the matrices *Transition* and *Emission* as *Parameters*. Our goal is to find *Parameters* and π when we are only given the emitted string x. We will work toward this goal by assuming that we are given x as well as either *Parameters* or π, and we must infer the remaining component. If x and *Parameters* are known, then we can find the most likely hidden path π using the Viterbi algorithm. However, we have not yet considered how to estimate *Parameters* if we know x and the hidden path π.

HMM Parameter Estimation Problem:

Find optimal parameters explaining the emitted string and the hidden path of an HMM.

Input: A string $x = x_1 \ldots x_n$ emitted by an HMM with unknown transition and emission probabilities following a known hidden path $\pi = \pi_1 \ldots - \pi_n$.

Output: A transition matrix *Transition* and an emission matrix *Emission* that maximize $\Pr(x, \pi)$ over all possible transition and emission matrices.

If we know both x and π, then we can compute empirical estimates for the transition and emission probabilities using a method similar to one we used for estimating parameters for profile HMMs. If $T_{l,k}$ denotes the number of transitions from state l to state k in the hidden path π, then we can estimate the probability *transition*$_{l,k}$ by computing the ratio of $T_{l,k}$ to the total number of transitions leaving state l,

$$transition_{l,k} = \frac{T_{l,k}}{\sum_{\text{all states } j} T_{l,j}}.$$

Likewise, if $E_k(b)$ denotes the number of times symbol b is emitted when the hidden path π is in state k, then we can estimate the probability $emission_k(b)$ as the ratio of $E_k(b)$ to the total number of emitted symbols from state k,

$$emission_k(b) = \frac{E_k(b)}{\sum_{\text{all symbols } c \text{ in the alphabet}} E_k(c)}.$$

It turns out that the above two formulas for computing *Transition* and *Emission* result in parameters solving the HMM Parameter Estimation Problem.

Viterbi learning

If we know x and *Parameters*, then we can construct the most likely path π by applying the Viterbi algorithm to solve the Decoding Problem:

$$(x, ?, Parameters) \rightarrow \pi$$

On the other hand, if we know x and π, then reconstructing *Parameters* amounts to solving the HMM Parameter Estimation Problem:

$$(x, \pi, ?) \rightarrow Parameters$$

STOP and Think: What do the expressions $(x, \pi, ?) \rightarrow$ *Parameters* and $(x, ?, Parameters) \rightarrow \pi$ remind you of?

HMM Parameter Learning Problem:
Estimate the parameters of an HMM explaining an emitted string.

Input: A string $x = x_1 \ldots x_n$ emitted by an HMM with unknown transition and emission probabilities.
Output: A transition matrix *Transition* and an emission matrix *Emission* that maximize $\Pr(x, \pi)$ over all possible transition and emission matrices and over all hidden paths π.

Unfortunately, the HMM Parameter Learning Problem is intractable, and so we will instead develop a heuristic that is analogous to the Lloyd algorithm for k-means clustering from Chapter 8. In that algorithm, illustrated in Figure 8.12, we iterated two steps, "From Centers to Clusters",

CHAPTER 10

$$(Data, ?, Centers) \to HiddenVector,$$

and "From Clusters to Centers",

$$(Data, HiddenVector, ?) \to Centers.$$

As for HMM parameter estimation, we begin with an initial random guess for *Parameters*. Then, we use the Viterbi algorithm to find the optimal hidden path π:

$$(x, ?, Parameters) \to \pi$$

Once we know π, we will question our original choice of *Parameters* and apply our solution to the HMM Parameter Estimation Problem to update *Parameters* based on x and π:

$$(x, \pi, ?) \to Parameters'$$

We then iterate over these two steps, hoping that the estimated parameters are getting closer and closer to the parameters solving the HMM Parameter Learning Problem:

$$\begin{aligned}(x, ?, Parameters) &\to (x, \pi, Parameters) \to (x, \pi, ?) \\ &\to (x, \pi, Parameters') \to (x, ?, Parameters') \\ &\to (x, \pi', Parameters') \to (x, \pi', ?) \\ &\to (x, \pi', Parameters'') \to \ldots\end{aligned}$$

This approach to learning the HMM's parameters is called **Viterbi learning**.

> **STOP and Think:** Can $\Pr(x, \pi)$ decrease during Viterbi learning? When would you decide to stop the Viterbi learning algorithm?

Note that we have not specified how Viterbi learning should terminate. In practice, there are various stopping rules to control its running time. For example, the algorithm can be stopped if the number of iterations exceeds a predefined threshold or if $\Pr(x, \pi)$ changes very little from one iteration to another.

Also, because Viterbi learning is dependent on the initial guess for *Parameters*, it may become stuck in a local optimum. Like other heuristics, it is often run many times, retaining the best choice of *Parameters*.

> **EXERCISE BREAK:** Apply Viterbi learning to learn parameters for an HMM modeling CG-islands as well as for the profile HMM for the gp120 HIV alignment in Figure 10.1.

Soft Decisions in Parameter Estimation

The Soft Decoding Problem

In Chapter 8, we introduced a "soft" clustering algorithm, based on the more general expectation maximization algorithm, that relaxed the Lloyd algorithm's rigid assignment of points to clusters. Analogously, by generating a single optimal hidden path, the Viterbi algorithm provides a rigid "yes" or "no" answer to the question of whether an HMM was in state k at time i. But how certain are we that this was the case?

Returning to the crooked casino analogy once more, say that the i-th coin flip is heads. If this flip occurs in the middle of ten consecutive heads, then you should be relatively confident that the biased coin was used. But what if, of the ten flips surrounding the i-th flip, six are heads and four are tails? In this case, you should be less certain that the biased coin was used.

In the case of an arbitrary HMM, we would like to compute the conditional probability $\Pr(\pi_i = k | x)$ that the HMM was in state k at time i given that it emitted string x.

Soft Decoding Problem:

Find the probability that an HMM was in a particular state at a particular moment given its emitted string.

Input: A string $x = x_1 \ldots x_n$ emitted by an HMM.
Output: The conditional probability $\Pr(\pi_i = k | x)$ that the HMM was in state k at step i given that it emitted x.

The unconditional probability that a hidden path will pass through state k at time i and emit x can be written as the sum

$$\Pr(\pi_i = k, x) = \sum_{\text{all paths } \pi \text{ with } \pi_i = k} \Pr(x, \pi).$$

The conditional probability $\Pr(\pi_i = k | x)$ is equal to the proportion of paths that pass through state k at time i and emit x with respect to all paths emitting x:

$$\begin{aligned}\Pr(\pi_i = k | x) &= \frac{\Pr(\pi_i = k, x)}{\Pr(x)} \\ &= \frac{\sum_{\text{all paths } \pi \text{ with } \pi_i = k} \Pr(x, \pi)}{\sum_{\text{all paths } \pi} \Pr(x, \pi)}.\end{aligned}$$

CHAPTER 10

STOP and Think: If the Viterbi algorithm for the crooked casino emits a path $\pi = \pi_1 \pi_2 \ldots \pi_n$ with $\pi_i = B$, is the dealer more likely to have used a biased coin at step i? Is it possible that $\pi_i = B$ but that $\Pr(\pi_i = B|x)$ is smaller than $\Pr(\pi_i = F|x)$?

The forward-backward algorithm

We note that $\Pr(\pi_i = k, x)$ is equal to the sum of product weights $\Pr(\pi, x)$ of all paths π through the Viterbi graph for x that pass through the node (k, i). As shown in Figure 10.24 (top), we can break each such path into a blue subpath from *source* to (k, i), which we denote π_{blue}, and a (red) subpath from (k, i) to *sink*, which we denote π_{red}. Writing WEIGHT(π_{blue}) and WEIGHT(π_{red}) as the respective product weights of these subpaths yields the recurrence

$$\Pr(\pi_i = k, x) = \sum_{\text{all paths } \pi \text{ with } \pi_i = k} \Pr(x, \pi)$$

$$= \sum_{\text{all paths } \pi_{\text{blue}}} \sum_{\text{all paths } \pi_{\text{red}}} \text{WEIGHT}(\pi_{\text{blue}}) \cdot \text{WEIGHT}(\pi_{\text{red}})$$

$$= \sum_{\text{all paths } \pi_{\text{blue}}} \text{WEIGHT}(\pi_{\text{blue}}) \cdot \sum_{\text{all paths } \pi_{\text{red}}} \text{WEIGHT}(\pi_{\text{red}}).$$

We have already computed the sum of product weights of all blue subpaths; it is just *forward*$_{k,i}$, which we encountered when solving the Outcome Likelihood Problem. Now we would like to compute the sum of product weights of all red subpaths, which we denote as *backward*$_{k,i}$, so that the preceding equation becomes

$$\Pr(\pi_i = k, x) = \textit{forward}_{k,i} \cdot \textit{backward}_{k,i}.$$

The name of *backward*$_{k,i}$ derives from the fact that to compute this value, we can simply *reverse* the directions of all edges in the Viterbi graph (Figure 10.24 (bottom)) and apply the same dynamic programming algorithm used to compute *forward*$_{k,i}$. Since the reversed edge connecting $(l, i+1)$ to (k, i) has weight $\text{WEIGHT}_i(k, l) = \textit{transition}_{k,l} \cdot \textit{emission}_l(x_{i+1})$, we have that

$$\textit{backward}_{k,i} = \sum_{\text{all states } l} \textit{backward}_{l,i+1} \cdot \text{WEIGHT}_i(k, l).$$

STOP and Think: How should this recurrence be initialized?

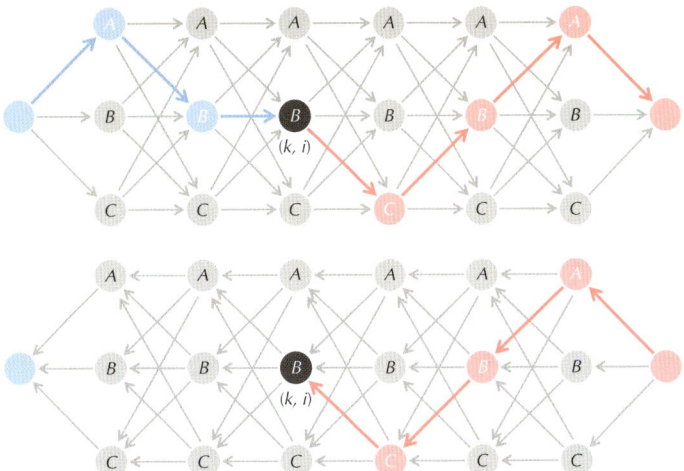

FIGURE 10.24 (Top) Each path from *source* to *sink* passing through the (black) node (k, i) in the Viterbi graph from Figure 10.7 (bottom) can be partitioned into two subpaths, one from *source* to (k, i) (shown in blue) and another from (k, i) to *sink* (shown in red). (Bottom) A "reversed Viterbi graph" in which all edges have been reversed, with a path from sink to (k, i) highlighted in red. The recurrence for $backward_{k,i}$ is based on computing $backward_{l,i+1}$ for every state l.

The resulting dynamic programming approach for computing $\Pr(\pi_i = k, x)$ is called the **forward-backward algorithm**. Combining the forward-backward algorithm with our solution to the Outcome Likelihood Problem for computing $\Pr(x)$ yields that

$$\Pr(\pi_i = k | x) = \frac{\Pr(\pi_i = k, x)}{\Pr(x)} = \frac{forward_{k,i} \cdot backward_{k,i}}{forward(sink)},$$

and so we are ready to solve the Soft Decoding Problem.

EXERCISE BREAK: Consider the following questions.

- For the crooked dealer HMM, compute $\Pr(\pi_i = k | x)$ for $x =$ "THTHH-HTHTTH" and each value of i. How does your answer change if $x =$ "HHHHHHHHHHH"?

- Apply your solution for the Soft Decoding Problem to find CG-islands in the first million nucleotides from the human X chromosome. How does your answer differ from the solution given by the Viterbi algorithm?

CHAPTER 10

We have just seen how to compute the conditional probability $\Pr(\pi_i = k | x)$ that the HMM passes through node (k, i) in the Viterbi graph given that the HMM emits x. But what about the conditional probability $\Pr(\pi_i = l, \pi_{i+1} = k | x)$ that the HMM passes through the edge connecting (l, i) to $(k, i+1)$ given that the HMM emits x? As with the forward-backward algorithm, we can divide every path through the edge in question into a blue path from *source* to this edge and a red path from this edge to *sink* (Figure 10.25).

EXERCISE BREAK: Prove that $\Pr(\pi_i = l, \pi_{i+1} = k | x)$ is equal to $forward_{l,i} \cdot \text{WEIGHT}_i(l, k) \cdot backward_{k,i+1} / forward(sink)$.

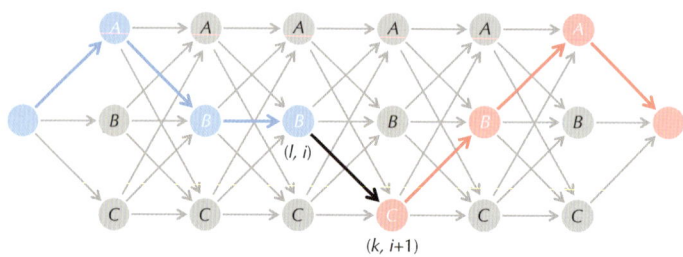

FIGURE 10.25 Each path from *source* to *sink* in the Viterbi graph passing through the (black) edge $(l, i) \to (k, i+1)$ in the Viterbi graph can be partitioned into two subpaths, one from *source* to (l, i) (shown in blue) and another from $(k, i+1)$ to *sink* (shown in red).

The probabilities $\Pr(\pi_i = k | x)$ can be put into a $|States| \times n$ responsibility matrix Π^*, where $\Pi^*_{k,i}$ corresponds to a *node* in the Viterbi graph and is equal to $\Pr(\pi_i = k | x)$. Figure 10.26 (top) shows the "responsibility" matrix Π^* for the crooked casino.

The probabilities $\Pr(\pi_i = l, \pi_{i+1} = k | x)$ can be put into another $|States| \times |States| \times (n-1)$ responsibility matrix Π^{**}, where $\Pi^{**}_{l,k,i}$ corresponds to an *edge* in the Viterbi graph and is equal to $\Pr(\pi_i = l, \pi_{i+1} = k | x)$ (Figure 10.26 (bottom)). For brevity, we use Π to collectively refer to the matrices Π^* and Π^{**}.

EXERCISE BREAK: What is the complexity of an algorithm computing the matrices Π^* and Π^{**}?

	T	H	T	H	H	H	T	H	T	T	H
F	0.636	0.593	0.600	0.533	0.515	0.544	0.627	0.633	0.692	0.686	0.609
B	0.364	0.407	0.400	0.467	0.485	0.456	0.373	0.367	0.308	0.314	0.391

	1	2	3	4	5	6	7	8	9	10
FF	0.562	0.548	0.507	0.473	0.478	0.523	0.582	0.608	0.643	0.588
FB	0.074	0.045	0.093	0.059	0.037	0.022	0.045	0.025	0.049	0.098
BF	0.031	0.053	0.025	0.042	0.066	0.104	0.051	0.084	0.043	0.022
BB	0.333	0.354	0.374	0.426	0.418	0.351	0.322	0.282	0.265	0.293

FIGURE 10.26 (Top) The responsibility matrix Π^*, where x = "THTHHHTHTTH" and the emission/transition matrices *Parameters* are taken from the crooked dealer HMM in Figure 10.5. $\Pi^*_{k,i}$ is equal to $\Pr(\pi_i = k|x)$. (Bottom) The responsibility matrix Π^{**}, where $\Pi^{**}_{l,k,i} = \Pr(\pi_i = l, \pi_{i+1} = k|x)$ for the same emitted string and emission/transition matrices.

Baum-Welch Learning

The expectation maximization algorithm for parameter estimation, called **Baum-Welch learning**, alternates between two steps. In the E-step, it estimates the responsibility profile Π given the current parameters:

$$(x, ?, Parameters) \to \Pi$$

Then, in the M-step, it re-estimates the parameters from the responsibility profile:

$$(x, \Pi, ?) \to Parameters$$

We have already implemented the E-step of the expectation maximization algorithm, but the question remains how to design the M-step.

When we know the hidden path, the previously defined estimators for *Parameters*, reproduced below, define optimal choices for a given hidden path π:

$$transition_{l,k} = \frac{T_{l,k}}{\sum_{\text{all states } j} T_{l,j}} \qquad emission_k(b) = \frac{E_k(b)}{\sum_{\text{all symbols } c \text{ in the alphabet}} E_k(c)}.$$

Here, $T_{l,k}$ is the number of transitions from state l to state k in the hidden path π, and $E_k(b)$ is the number of times symbol b is emitted when the hidden path π is in state k.

EXERCISE BREAK: How would you redefine these estimators when the hidden path is unknown?

To see how to estimate *transition*$_{l,k}$ and *emission*$_k(b)$ when the hidden path is unknown, we will compute $T_{l,k}$ and $E_k(b)$ for a known path π in a slightly different way to make the transition from hard to soft choices more apparent. First, define the following binary variables:

$$T_{l,k}^i = \begin{cases} 1 \text{ if } \pi_i = l \text{ and } \pi_{i+1} = k \\ 0 \text{ otherwise} \end{cases} \qquad E_k^i(b) = \begin{cases} 1 \text{ if } \pi_i = k \text{ and } x_i = b \\ 0 \text{ otherwise} \end{cases}$$

With this notation, the formulas computing $T_{l,k}$ and $E_k(b)$ can be rewritten as

$$T_{l,k} = \sum_{i=1}^{n-1} T_{l,k}^i \qquad E_k(b) = \sum_{i=1}^{n} E_k^i(b)$$

When the hidden path is unknown, we will substitute the binary variables $T_{l,k}$ and $E_k(b)$ for new variables $T_{l,k}^i$ and $E_k^i(b)$ that are computed in terms of the conditional probabilities that a hidden path will pass through a given node or edge of the Viterbi graph:

$$T_{l,k}^i = \Pr(\pi_i = l, \pi_{i+1} = k | x) \qquad E_k^i(b) = \Pr(\pi_i = k | x)$$
$$= \Pi_{l,k,i}^{**} \qquad\qquad\qquad = \Pi_{k,i}^* \text{ if } x_i = b \text{ and } 0 \text{ otherwise}$$

Armed with these probabilities computed in the previous section, we can compute new estimates for *Parameters* that often perform better in practice than the estimates provided by Viterbi learning:

$$transition_{l,k} = \sum_{i=1}^{n-1} \Pi_{l,k,i}^{**} \qquad emission_k(b) = \sum_{i=1}^{n} \Pi_{k,i}^*$$

STOP and Think: Should we normalize the transition and emission probabilities in the above equations? For example, do the above equations imply that all transition probabilities leaving state l must sum to 1?

EXERCISE BREAK: Use Baum-Welch learning to learn parameters for the HMM modeling CG-islands and for the HIV profile HMM. Compare these parameters with parameters derived by applying Viterbi learning.

The Many Faces of HMMs

Profile HMMs for multiple sequence alignment and HMMs finding CG-islands are just two examples of many applications of HMM in bioinformatics. Furthermore, applications of HMMs to HIV analysis are not limited to profile HMMs but also include analysis of HIV resistance against antiviral drug therapies.

Early in the chapter, we mentioned that patients infected with HIV are treated with a cocktail of several drugs. These drugs attempt to suppress replication of the virus, but HIV often mutates into drug-resistant strains that eventually dominate the virus population in a host, making a drug cocktail progressively ineffective. HIV viruses are often sequenced after drug therapy has failed in order to decide how to reformulate the drug cocktail. Thus, understanding HIV's pathways to drug resistance is important to design an effective cocktail.

Yet modeling HIV resistance pathways is a difficult task. Two mutations that are advantageous for the virus may interact synergistically, causing the double mutation to be fixed more often than we might predict from the frequencies of the individual substitutions. Mutations can also interact antagonistically, resulting in mutants that are less fit than we might predict.

In 2007, Niko Beerenwinkel and Mathias Drton introduced an HMM-based model for HIV evolution and developing drug resistance. However, their HMM is far too complex to explain here. We nevertheless mention it here in order to emphasize the power of HMMs. Even though they may seem like simple machines that flip coins and emit symbols, HMMs can be applied to tackle complex bioinformatics problems ranging from gene prediction to regulatory motif finding.

Epilogue: Nature is a Tinkerer and not an Inventor

The sequence of amino acids in a protein encodes its 3-D structure, which often defines the biological function of a protein. For example, a **zinc finger** is an element of the 3-D structure of **zinc finger proteins** (Figure 10.27). By arranging two cysteines and two histidines close to each other in a zinc finger protein's amino acid sequence, the protein is able to "grab" a zinc ion and fold tightly around it. Zinc fingers are so useful that they are found in thousands of human proteins. Furthermore, zinc finger proteins are used for more than binding zinc, as many of these proteins bind to other metals or even non-metals.

Over 100,000 experimentally determined protein structures are currently known, but many of them represent very similar structures or share segments with very similar

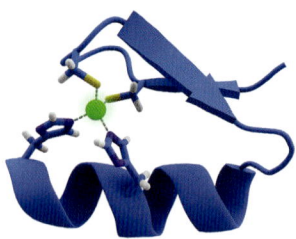

FIGURE 10.27 A zinc ion (shown in green) is held in place by two histidine residues and two cysteine residues within a zinc finger.

structure. A **protein domain** is a conserved part of a protein that can function independently of the rest of the protein. Domains vary in length, but the average domain length is approximately 100 amino acids (zinc finger domains are only 20-30 amino acids long.) Many proteins consist of several domains, and the same domain may appear (with variations) in many proteins.

Nobel Laureate François Jacob famously said in 1977: "Nature is a tinkerer and not an inventor." In accordance with this principle, nature uses domains as building blocks, shuffling them into different arrangements to create **multi-domain proteins**. Most domains once existed as independent proteins; for example, many domains belonging to human multi-domain proteins can be found as single-domain proteins in bacteria. Multi-domain proteins occur naturally when a genome rearrangement creates a new protein-coding sequence containing parts of the coding sequences from two different genes. Association of two domains into a single protein often provides an evolutionary advantage, such as when both domains are enzymes, in which case it may be beneficial for the cell to ensure a fixed one-to-one ratio of the enzymes' activities.

Since proteins are often built from multiple domains with different structures and functions, biologists commonly analyze individual domains instead of entire proteins in order to understand evolutionary relationships. Since sequence similarities between domains with similar structures can be extremely low, classifying domains into structural families can be difficult. The **Pfam database**, which contains over 10,000 HMM-derived multiple alignments of protein domain families, can be used to analyze new protein sequences.

> **CHALLENGE PROBLEM:** Using the Pfam HMM for gp120 (constructed from a seed alignment of just 24 gp120 proteins), construct alignments of all known gp120 proteins and identify the "most diverged" gp120 sequence.

Detours

The Red Queen Effect

The **Red Queen Effect** is the hypothesis that evolution is necessary not only to equip organisms with an advantage in a *fixed* environment, but also to help them survive in response to *changing* environments. Its name derives from a statement that the Red Queen made to Alice in Lewis Carroll's *Through the Looking-Glass*:

> *Now, here, you see, it takes all the running you can do, to keep in the same place.*

The Red Queen Effect is often seen in predator-prey relationships. For example, an adaptation may help wolves run a little faster, and caribou must evolve in turn to survive. The result is that wolves and caribou appear to run at the same speed, with the slowest wolves are starving and the slowest caribou being eaten.

Glycosylation

Cells have a dense coating of sugar chains, called **glycans**, on their surface. Glycans are often post-translational modifications of **glycoproteins**, which modulate interactions with other cells in a multicellular organism or between the cell and another organism (e.g., between human cells and a virus). For example, influenza infection begins with an interaction between the proteins on the virus's surface and glycans on the host cell's surface.

Glycans are constructed from a family of building blocks called **monosaccharides**. Each monosaccharide can be linked with other monosaccharides to form complex, tree-like structures (Figure 10.28).

DNA methylation

DNA methylation results in the addition of a methyl group (CH_3) to a cytosine or guanine nucleotide (Figure 10.29), which often alters the expression of nearby genes. Genes that acquire a high concentration of methylated residues in their upstream regions have suppressed expression. DNA methylation is vital to development, and both DNA hypermethylation and hypomethylation have been linked to various cancers.

DNA methylation is important in the process of **cell differentiation**, in which embryonic stem cells become specialized tissues. The change is often permanent, preventing a cell from reverting to a stem cell or converting to a different cell type. Methylation is inherited during cell division but is usually removed during zygote formation.

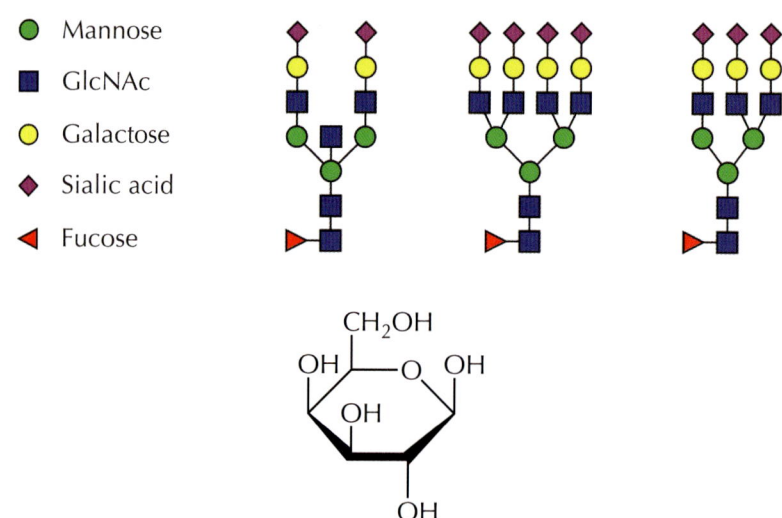

FIGURE 10.28 (Top) Five types of monosaccharides along with three examples of how these monosaccharides are assembled into glycans in humans. (Bottom) The chemical formula for the monosaccharide galactose.

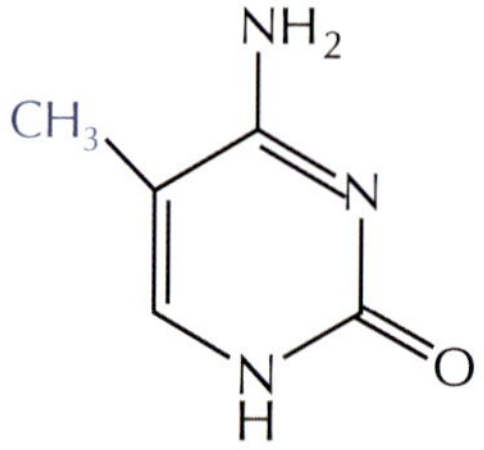

FIGURE 10.29 DNA methylation of a cytosine nucleotide base, with the methyl group shown in blue.

Conditional probability

Let's return to the game of Chō-Han and analyze the sum s of two standard six-sided dice. Let A be the event that s is odd and B be the event that s is larger than 10. The probability of A is equal to 1/2 because half of the 36 possible outcomes for rolling two dice produce an odd sum. The probability of B is equal to 3/36 because there are three outcomes $(5 + 6, 6 + 5,$ and $6 + 6)$ for which $s > 10$.

STOP and Think: If we tell you that s is larger than 10 (but you cannot see the dice), is s more likely to be odd or even?

The **conditional probability** of event A given event B, denoted $\Pr(A|B)$, is the probability that event A will occur given that event B has occurred. For the dice example, since B corresponds to two tosses with s odd ($6+5$ and $5+6$) and one toss with s even ($6+6$), $\Pr(A|B) = 2/3$. Note that $\Pr(A|B)$ is completely different from the probability $\Pr(A, B)$ that both events A and B will occur, which is equal to $2/36$ (A and B only occur together for the sums $6+5$ and $5+6$).

More generally, the conditional probability $\Pr(A|B)$ is often defined using the following formula:

$$\Pr(A|B) = \frac{\Pr(A,B)}{\Pr(B)}.$$

EXERCISE BREAK: To test your knowledge of conditional probability, consider the following puzzle, called the "Monty Hall Problem", which originally appeared in a letter to *American Statistician* in 1975 and has stumped many aspiring mathematicians over the years:

> *Suppose you're on a game show, and you're given the choice of three doors: Behind one door is a car; behind the others, goats. You pick a door, say No. 1, and the host, who knows what's behind the doors, opens another door, say No. 3, which has a goat. He then says to you, "Do you want to pick door No. 2?" Is it to your advantage to switch your choice?*

Bibliography Notes

The decoding algorithm was developed by Viterbi, 1967. The first algorithms for HMM parameter estimation were developed by Baum et al., 1970. Churchill, 1989, Krogh et al., 1994, and Baldi et al., 1994 pioneered the application of HMMs in computational biology. Bateman et al., 2002 described applications of profile HMM alignments for developing a database of protein domain families Pfam. De Jong et al., 1992 discovered the 11/25 rule. Beerenwinkel and Drton, 2007 constructed an HMM for analyzing HIV resistance.

WHY HAVE BIOLOGISTS STILL NOT DEVELOPED AN HIV VACCINE?

Paleontology Meets Computing

Growing up in Montana in the 1950s, Jack Horner was shy and introverted. He progressed so slowly in reading and mathematics that other kids called him stupid. However, his high school project on dinosaurs took top honors at the local science fair and was noticed by a University of Montana professor, who helped Jack gain admission to the university.

Yet Horner's grades did not improve in college; after failing five consecutive quarters, he dropped out. Years later, Horner would learn that he suffers from **dyslexia**, a developmental disorder that is often characterized by difficulty with reading comprehension and mathematics despite normal or above-average intelligence.

Fortunately for Horner, he eventually found his calling. After being drafted during the Vietnam War and later working as a truck driver, he accepted a job as a technician at Princeton's Natural History Museum, where he quickly established a reputation among his peers as a brilliant researcher. He would go on to become the world's most famous paleontologist, providing inspiration for one of the main characters of the bestselling novel *Jurassic Park* and advising Steven Spielberg for the film adaptation.

Horner was able to succeed despite dyslexia partly because paleontology has not traditionally required mathematical fluency. However, Horner's own student would show that even paleontology is not immune from computing. In 2000, Horner was exploring his favorite dinosaur graveyard in Montana and discovered a 68 million year-old *Tyrannosaurus rex* leg bone fossil. Three years later, he gave a small chunk of this fossil to his student, Mary Schweitzer, who dissolved it in a demineralizing bath to study its components but left it in for too long (remember Alexander Fleming?). When she returned, all that remained was a fibrous substance. Schweitzer then sent this material to a mass spectrometrist (John Asara) in the hope of detecting *T. rex* peptides, or short protein fragments, which had miraculously survived inside the bone.

In 2007, after analyzing thousands of spectra, Asara and Schweitzer published a paper in *Science* announcing the discovery of *T. rex* peptides that closely matched chicken peptides. Their result provided the first molecular evidence for the controversial hypothesis that birds evolved from dinosaurs.

The fact that proteins could survive for millions of years was so amazing that it led to many grandiose claims. Paleontologist Hans Larsson suggested that dinosaurs would "enter the field of molecular biology and really slingshot paleontology into the modern world". *The Guardian* projected that "scientists may one day be able to emulate *Jurassic Park* by cloning a dinosaur." Horner himself even published a book called *How*

CHAPTER 11

to Build a Dinosaur, detailing his plan to recreate a dinosaur by genetically modifying the chicken genome.

Yet some scientists remained skeptical. Whereas previous dinosaur studies did not require much computation, Asara's *T. rex* analysis was powered by an algorithm relying on complicated statistics. In 2008, *Science* published rebuttals arguing that Asara and Schweitzer had failed to prove that some of their peptides are not simply statistical artifacts. But how can we know which side is correct? In this chapter, we will investigate the *T. rex* peptides by delving into some algorithms for analyzing spectra.

Which Proteins Are Present in This Sample?

Only four scientists have ever won two Nobel Prizes. One of them is Frederick Sanger, whose assembly of the first genome in 1977 we mentioned in Chapter 3. Yet Sanger had already won his first Nobel prize two decades earlier for determining the sequence of 52 amino acids making up insulin, the protein needed to absorb glucose in the blood. Similarly to how scientists sequence genomes, Sanger broke multiple molecules of insulin into short peptides, sequenced these peptides, and then assembled them into the amino acid sequence of insulin (Figure 11.1).

Although protein sequencing was very difficult in the 1950s, DNA sequencing was impossible. Today, it has become essentially trivial to generate millions of reads for DNA sequencing, but protein sequencing remains difficult. For this reason, most proteins are discovered by first sequencing a genome and then predicting all of the genes that this genome encodes (see **DETOUR: Gene Prediction**). By translating the nucleotide sequence of each protein-coding gene into an amino acid sequence, biologists derive a putative **proteome** of a species, i.e., the set of all its proteins.

However, different cells in an organism express different proteins. For example, brain cells express proteins giving rise to neuropeptides, whereas other cells do not. An important problem in the study of proteins, or **proteomics**, is to identify which *specific* proteins are present in each biological tissue under different conditions, and how these proteins interact.

For example, suppose we are studying the chicken ribosome, a complex molecular machine consisting of many proteins. Knowing the chicken proteome does not tell us which specific proteins compose the ribosome complex. Instead, we can isolate the ribosome, break it apart, and identify which proteins it contains. In practice, merely confirming that a 10 amino acid-long peptide from a known chicken protein is present in a sample is usually sufficient to confirm this protein's presence in the sample. The

```
GIVEECCA
GIVEECCASV
GIVEECCASVC
GIVEECCASVCSL
GIVEECCASVCSLY
            SLYELEDYC
                ELEDY
                ELEDYCD
                 LEDYCD
                  EDYCD
                      FVDEHLCG
                      FVDEHLCGSHL
                          HLCGSHL
                             SHLVEA
                                VEALY
                                  YLVCG
                                   LVCGERGF
                                   LVCGERGFF
                                        GFFYTPK
                                         YTPKA
```
GIVECCASVCSLYELEDYCDFVDEHLCGSHLVEALYLVCGERGFFFYTPKA

FIGURE 11.1 The peptide assembly that Frederick Sanger used to determine the amino acid sequence of insulin.

process of confirming that a peptide from a *known* proteome is present in a sample is called **peptide identification**. But how could we form a *T. rex* proteome?

Although peptide identification dominates modern proteomics studies, the proteomes of many species, including extinct species like *T. rex*, remain unknown. In this case, biologists rely on *de novo* **peptide sequencing**, or inferring the amino acid sequence of a peptide without relying on a proteome, and so this is where we will begin.

Decoding an Ideal Spectrum

You may be experiencing *déjà vu*, since we already discussed *cyclic* peptide sequencing in Chapter 4. So we will first remind you of the basics of mass spectrometry, with an emphasis on its application to *linear* peptide sequencing.

Given a large number of identical copies of a peptide in a sample — which will typically contain millions of cells — a mass spectrometer breaks each copy into two smaller fragments, where different copies of the same peptide may break differently. For example, one copy of REDCA may break into RE and DCA, and another may break into

RED and CA. The fragments RE and RED are called **prefixes** of REDCA, whereas DCA and CA are called **suffixes** of REDCA. Figure 11.2 shows the integer masses of amino acids.

G	A	S	P	V	T	C	I	L	N	D	K	Q	E	M	H	F	R	Y	W
57	71	87	97	99	101	103	113	113	114	115	128	128	129	131	137	147	156	163	186

FIGURE 11.2 The integer mass table of the 20 standard amino acids (reproduced from Chapter 4).

The algorithmic question that we first pose is analogous to the one that we asked about cyclic antibiotics but now applied to linear peptides: if we weigh each prefix and suffix of an unknown peptide, can we reconstruct the peptide? Given an amino acid string *Peptide*, its ideal spectrum, denoted IDEALSPECTRUM(*Peptide*), is the collection of integer masses of all its prefixes and suffixes (Figure 11.3 (top)). Note that an ideal spectrum may have repeated masses; for example, IDEALSPECTRUM(GPG) = {0, 57, 57, 154, 154, 211}. We say that an amino acid string *Peptide* **explains** a collection of integers *Spectrum* if IDEALSPECTRUM(*Peptide*) = *Spectrum*.

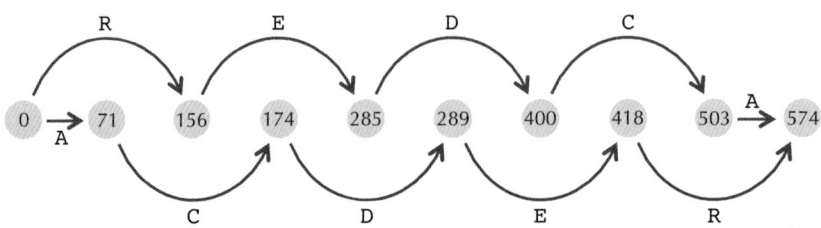

FIGURE 11.3 (Top) Masses of prefixes and suffixes of REDCA form IDEALSPECTRUM(REDCA) = {0, 71, 156, 174, 285, 289, 400, 418, 503, 574}. (Bottom) The DAG GRAPH(IDEALSPECTRUM(REDCA)) in which each mass in the spectrum is assigned to a node and two nodes are connected with a directed edge if the difference in their masses is the mass of an amino acid.

Decoding an Ideal Spectrum Problem:
Reconstruct a peptide from its ideal spectrum.

Input: A collection of integers *Spectrum*.
Output: An amino acid string *Peptide* that explains *Spectrum*.

We would like to separate masses in a spectrum into those derived from prefix and suffix peptides, but it is unclear how to do so. Instead, note that if two masses are "one amino acid mass apart", then it is likely that they correspond to two prefixes or two suffixes that differ by a single amino acid. For example, we would not know that masses 400 and 503 correspond to the prefixes RED and REDC (Figure 11.3 (top)). But we could hypothesize that because the difference between these masses is 103 (the mass of C), these masses correspond to prefixes or suffixes differing in a single occurrence of C.

This idea motivates a graph-based approach to solving the Decoding an Ideal Spectrum Problem. We represent the masses in a spectrum as a sequence *Spectrum* of integers $s_1, \ldots, s_m$ in increasing order, where s_1 is zero and s_m is the total mass of the (unknown) peptide. We define a labeled graph GRAPH(*Spectrum*) by forming a node for each element of *Spectrum*, then connecting nodes s_i and s_j by a directed edge labeled by an amino acid a if $s_j - s_i$ is equal to the mass of a (Figure 11.3 (bottom)). As we assumed when sequencing antibiotics, we do not distinguish between amino acids having the same integer masses (i.e., the pairs K/Q and I/L).

> **EXERCISE BREAK:** Prove that for any choice of *Spectrum*, GRAPH(*Spectrum*) is a DAG.

Figure 11.3 (bottom) shows that GRAPH(*IdealSpectrum*(REDCA)) consists of two paths connecting *source* = 0 to *sink* = s_m. Concatenating the amino acids along these paths spells out REDCA and its reverse ACDER, both of which represent solutions of the Decoding an Ideal Spectrum Problem. The sequencing approach based on spelling a path from *source* to *sink* in GRAPH(*Spectrum*) is described by the following pseudocode.

DECODINGIDEALSPECTRUM(*Spectrum*)
　construct GRAPH(*Spectrum*)
　find a path *Path* from *source* to *sink* in GRAPH(*Spectrum*)
　return the amino acid string spelled by labels of *Path*

CHAPTER 11

EXERCISE BREAK: Decode the ideal spectrum {0, 57, 114, 128, 215, 229, 316, 330, 387, 444}.

If you attempted this exercise, then there is a good chance that you found a path corresponding to a peptide with an incorrect spectrum (Figure 11.4). It is true that each peptide explaining *Spectrum* corresponds to a path from *source* to *sink* in GRAPH(*Spectrum*). However, not every path from *source* to *sink* in this graph corresponds to a peptide explaining *Spectrum*; consider, for example, the path spelling out GGDTN in Figure 11.4. For this reason, we must rewrite the above faulty pseudocode as follows.

DECODINGIDEALSPECTRUM(*Spectrum*)
    construct GRAPH(*Spectrum*)
    for each path *Path* from *source* to *sink* in GRAPH(*Spectrum*)
        Peptide ← the amino acid string spelled by the edge labels of *Path*
        if IDEALSPECTRUM(*Peptide*) = *Spectrum*
            return *Peptide*

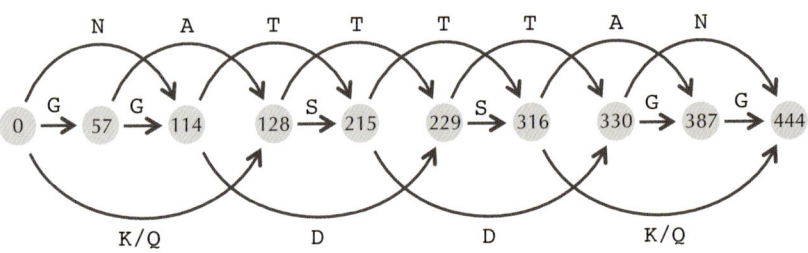

FIGURE 11.4 The DAG GRAPH(*Spectrum*) for *Spectrum* = {0, 57, 114, 128, 215, 229, 316, 330, 387, 444}. Only eight of the 32 paths from *source* to *sink* in this graph correspond to peptides explaining *Spectrum*).

Although **DECODINGIDEALSPECTRUM** solves the Decoding an Ideal Spectrum Problem, exploring all paths in a DAG may be time-consuming, since the number of such paths may be exponential in the number of masses in the spectrum (see **DETOUR: Finding All Paths in a Graph**).

From Ideal to Real Spectra

We already know from our analysis of antibiotics in Chapter 4 that the realities of peptide sequencing are more harsh than reconstructing a peptide from an ideal spectrum. After breaking each copy of a peptide into two smaller fragments, a mass spectrometer ionizes them, resulting in electrically charged **fragment ions**. It measures each fragment ion's **mass-to-charge ratio** as well as its **intensity**, or the number of fragment ions detected at that mass-to-charge ratio (peptides may break frequently at some bonds and hardly ever at other bonds). As a result, a spectrum is represented as a collection of peaks in a chart, where a peak's x-coordinate represents its mass-to-charge ratio, and its height represents its intensity (Figure 11.5 (top)).

Modern mass spectrometers have limitations on the range of mass-to-charge ratios that they can detect, making it difficult to analyze entire proteins by mass spectrometry. As a result, proteins are usually analyzed by first breaking them into shorter peptides using enzymes called **proteases**. The most popular protease used in proteomics, and the one used in the *T. rex* study, is called **trypsin**. This protease typically breaks a protein after the amino acids R and K and results in peptides of average length 14.

Figure 11.5 shows a mass spectrum for one of the *T. rex* spectra (henceforth referred to as *DinosaurSpectrum*), along with its two putative interpretations, ATKIVDCFMTY and GLVGAPGLRGLPGK. Once we infer the peptide that generated a given spectrum, we can **annotate** the spectrum by establishing a correspondence between peaks in the spectrum and prefixes/suffixes of the peptide. To comply with standard mass spectrometry terminology, a peak annotated as the prefix of length i is labeled b_i, and a peak annotated as the suffix of length i is labeled y_i.

> **EXERCISE BREAK:** Which of the two interpretations of *DinosaurSpectrum* in Figure 11.6 do you think better explains *DinosaurSpectrum*?

Sequencing a peptide from its real spectrum is even more difficult than it may already seem. Mass spectra often have "noisy" peaks that contribute to false masses, which may have higher intensities than peaks corresponding to true prefixes and suffixes. Since some peptide bonds hardly ever break, the intensities at different mass-to-charge ratios may differ by orders of magnitude. As a result, a spectrum may not have peaks corresponding to true prefixes and suffixes. For example, in Figure 11.5 (bottom), there are no peaks annotated as b_5 or y_9 in *DinosaurSpectrum*. For this reason, although we ignored intensities when sequencing antibiotics, we will take them more seriously in this chapter.

CHAPTER 11

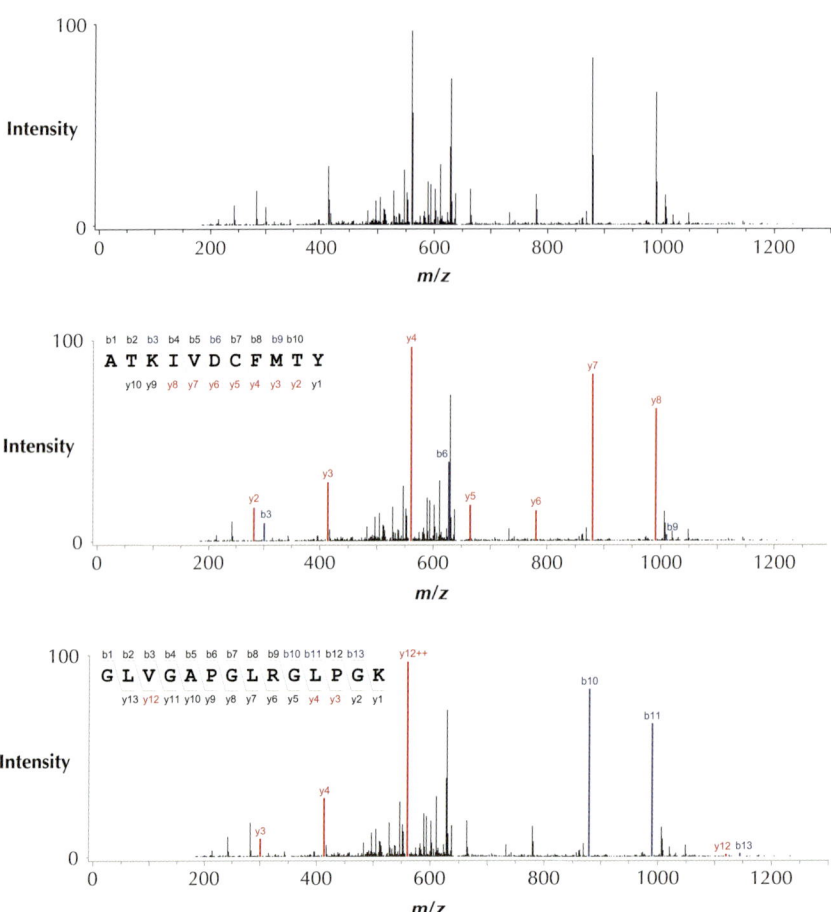

FIGURE 11.5 (Top) The spectrum *DinosaurSpectrum*. (Middle) The same spectrum annotated by ATKIVDCFMTY. (Bottom) The same spectrum annotated by GLVGAPGLRGLPGK. The peak corresponding to the prefix peptide of length *i* is annotated as b_i, and the peak corresponding to the suffix peptide of length *i* is annotated as y_i. For example, the peak annotated as b_{10} corresponds to GLVGAPGLRG, and the peak annotated as y_3 corresponds to PGK. Most annotated peaks have charge +1, but some (such as the one denoted y_{12}++) have charge +2. The charge of the fragment ion represented by a given peak in the spectrum is not known in advance but can often be inferred after the peptide that generated the spectrum has been identified. Only six peaks in *DinosaurSpectrum* are annotated by GLVGAPGLRGLPGK; peaks b_{10}, b_{11}, and b_{13} are annotated by prefix peptides, and peaks y_3, y_4, and y_{12} are annotated by suffix peptides.

EXERCISE BREAK: REDCA is one possible peptide that explains some but not all of the masses in the spectrum shown in Figure 11.6, which has false and missing masses. Can you find another peptide that explains even more masses in this spectrum?

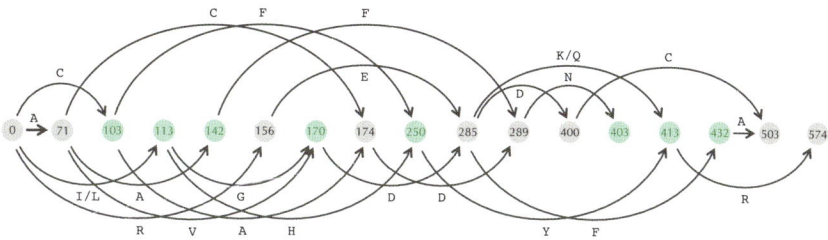

FIGURE 11.6 The DAG GRAPH(*Spectrum*) constructed from *Spectrum* = {0, 71, 103, 113, 142, 156, 170, 174, 250, 285, 289, 400, 403, 413, 432, 503, 574} with one missing mass (418) and eight false masses (shown in green) compared to the ideal spectrum of REDCA.

The issue of false and missing masses is only one of the many complications in mass spectrometry. When a mass spectrometer breaks a peptide, small parts of the resulting fragments may be lost, thus lowering their mass. For example, when breaking REDCA into RE and DCA, RE might lose a water molecule (H_2O) of mass $1 + 1 + 16 = 18$, and DCA might lose an ammonia molecule (NH_3) of mass $1 + 1 + 1 + 14 = 17$. The respective integer masses of the resulting fragments will be equal to MASS(RE) − 18 and MASS(DCA) − 17.

Because of the many practical complications of mass spectrometry, we will need to make some simplifying assumptions in order to move toward a computational problem modeling peptide sequencing. Instead of trying to account for the great variety of different fragmentation patterns when reconstructing a peptide, we will treat them as noise. We will also assume that all peaks have charge +1 and that the spectra are discretized (i.e., that all masses are integers).

CHAPTER 11

Peptide Sequencing

Scoring peptides against spectra

Consider an imaginary world in which peptides are built from just two amino acids, X and Z, having respective masses 4 and 5. For example, given the peptide XZZXX, prefixes have masses 4, 9, 14, 18, and 22, while suffixes have masses 22, 18, 13, 8, 4.

Now consider the hypothetical spectrum

$$(0, 0, 0, 3, 8, 7, 2, 1, 100, 0, 1, 4, 3, 500, 2, 1, 3, 9, 1, 2, 2, 0)$$

arising from this peptide, where the *i*-th element of this vector corresponds to the intensity detected at mass *i*. The prefixes of XZZXX annotate peaks with intensities 3, 100, 500, 9, and 0, whereas the suffixes annotate peaks with intensities 0, 8, 0, 2, and 1. Our goal is to develop an approach for scoring peptides against spectra in the hope that we will be able to find a peptide that generated the spectrum by simply finding the peptide with maximum score against this spectrum.

> **STOP and Think:** How would you score a peptide against a spectrum?

One scoring approach is the **intensity count**, or the sum of intensities of all peaks annotated by a peptide. For example, we would score peptide XZZXX against the spectrum above as the sum of intensities of all peaks annotated by XZZXX, i.e., $3 + 100 + 500 + 9 + 0 + 0 + 8 + 0 + 2 + 1$. However, the intensity count does not work well in practice because peak intensities vary widely. As a result, the tallest peaks in the spectrum (100 and 500 in our toy example) may dominate the score. Since the highest peaks may represent noise and since lower intensity peaks often represent correct prefix/suffix peptides, the intensity count is not a good scoring function in practice.

Another scoring approach, called the **shared peaks count**, simply counts the number of "tall" peaks annotated by a peptide, i.e., the annotated peaks with intensities exceeding a predefined threshold. Taking the intensity threshold 5 in our ongoing example, the shared peak count is 4, since the prefixes of XZZXX annotate the tall peaks with intensities 100, 500, and 9, and the suffixes annotate the tall peak with intensity 8. The peptide in Figure 11.5 (middle) has shared peak count equal to 10, whereas the peptide in Figure 11.5 (bottom) has shared peak count equal to 6.

Although the shared peak count works better than the intensity count in practice, it is still far from ideal; a better approach would account for intensities of peaks without letting the tallest peaks dominate the score. To achieve this goal, we will convert

WAS T. REX JUST A BIG CHICKEN?

peptides and spectra into vectors, then define a scoring function that is the dot product of these vectors.

First, given an amino acid string $Peptide = a_1 \ldots a_n$ of length n, we will represent its prefix masses using a binary **peptide vector** $\overrightarrow{Peptide}$ with $\text{MASS}(Peptide)$ coordinates. This vector contains a 1 at each of the n **prefix coordinates**

$$\text{MASS}(a_1), \text{MASS}(a_1 a_2), \ldots, \text{MASS}(a_1 a_2 \ldots a_n),$$

and it contains a 0 in each of the remaining noise coordinates. The toy peptide XZZXX, whose prefix masses are 4, 9, 14, 18, and 22, corresponds to the peptide vector $(0, 0, 0, 1, 0, 0, 0, 0, 1, 0, 0, 0, 0, 1, 0, 0, 0, 1, 0, 0, 0, 1)$ of length 22 (Figure 11.7).

	1	2	3	4	5	6	7	8	9	10	11	12	13	14	15	16	17	18	19	20	21	22
peptide vector	0	0	0	**1**	0	0	0	0	**1**	0	0	0	0	**1**	0	0	0	**1**	0	0	0	**1**
spectral vector	0	0	0	4	-2	-3	-1	-7	6	5	3	2	1	9	3	-8	0	3	1	2	1	0

FIGURE 11.7 The peptide vector of XZZXX and a hypothetical spectral vector generated by this peptide (assuming that X and Z have masses 4 and 5, respectively). Prefix coordinates of the peptide vector are shown in boldface. Amplitudes exceeding the threshold 3 in the spectral vector are shown in color. These bold entries correspond to three prefix coordinates (blue) and one noise coordinate (red).

Converting a Peptide into a Peptide Vector Problem:

Convert a peptide into a peptide vector.

> **Input**: An amino acid string *Peptide*.
> **Output**: The peptide vector $\overrightarrow{Peptide}$.

Since a peptide vector uniquely defines the peptide that it originated from, we will use the terms "peptide vector" and "peptide" interchangeably.

Converting a Peptide Vector into a Peptide Problem:

Convert a peptide vector into a peptide.

> **Input**: A binary vector P.
> **Output**: A peptide whose peptide vector is equal to P (if such a peptide exists).

CHAPTER 11

Where are the suffix peptides?

You may be wondering why peptide vectors only model prefix peptides, since both prefix and suffix peptides contribute to spectral annotations. Does it not make more sense to define the peptide vector of XZZXX as (0, 0, 0, 1, 0, 0, 0, 1, 1, 0, 0, 0, 1, 1, 0, 0, 0, 1, 0, 0, 0, 1) in order to reflect both its prefix masses (4, 9, 14, 18, and 22) and suffix masses (4, 8, 13, 18, 22)?

Indeed, any peak of mass s in a spectrum may be interpreted as either a prefix mass or suffix mass of an unknown peptide *Peptide* that generated the spectrum. Moreover, its **twin peak**, with mass equal to MASS(*Peptide*) - s, may be interpreted as either a mass of a suffix or a mass of a prefix of the same peptide.

To deal with this uncertainty, mass spectrometrists convert a spectrum *Spectrum* into a **spectral vector** $\overrightarrow{Spectrum}$ that consolidates information about the intensities of each peak and its twin into a single value, called an **amplitude**, at the coordinate representing the mass of the hypothetical prefix peptide for these twins (Figure 11.8). Why? Because algorithms interpreting a consolidated spectrum become easier than algorithms attempting to account for both twins (see **DETOUR: The Anti-Symmetric Path Problem**). To make matters more complicated, amplitudes in the spectral vectors also account for intensities of ion fragments with various charges and intensities of ion fragments with water and ammonia molecule loss.

Figure 11.8 shows a spectral vector constructed from *DinosaurSpectrum* and illustrates that amplitudes in a spectral vector may be negative. Negative amplitudes typically correspond to positions in spectra without peaks or with low intensity peaks. The correspondence between intensities in a spectrum and amplitudes in its spectral vector is complex, but in general, the amplitude at mass i reflects the *likelihood* that the (unknown) peptide that generated the spectrum has a prefix with mass i (see **DETOUR: Transforming Spectra into Spectral Vectors**).

After a peptide *Peptide* has been transformed into a peptide vector $\overrightarrow{Peptide} = (p_1, \ldots, p_m)$ and a spectrum *Spectrum* has been transformed into a spectral vector $\overrightarrow{Spectrum} = (s_1, \ldots, s_m)$ of the same length, we define SCORE(*Peptide*, *Spectrum*) = SCORE($\overrightarrow{Peptide}$, $\overrightarrow{Spectrum}$) as the dot product of $\overrightarrow{Peptide}$ and $\overrightarrow{Spectrum}$,

$$\text{SCORE}(Peptide, Spectrum) = p_1 \cdot s_1 + \ldots + p_m \cdot s_m.$$

Note that SCORE($\overrightarrow{Peptide}$, $\overrightarrow{Spectrum}$) is simply an "amplitude count", the sum of amplitudes in $\overrightarrow{Spectrum}$ that are "annotated" by $\overrightarrow{Peptide}$. However, this score does not suffer from the limitations of the intensity count because we have transformed intensities in a spectrum into amplitudes in its spectral vector. As a result, high-intensity peaks in the spectrum contribute to the score but do not dominate it. The score of the peptide vector and spectral vector in Figure 11.7 is $4 + 6 + 9 + 3 + 0 = 22$; the score of the peptide vector and spectral vector in Figure 11.8 is -19.

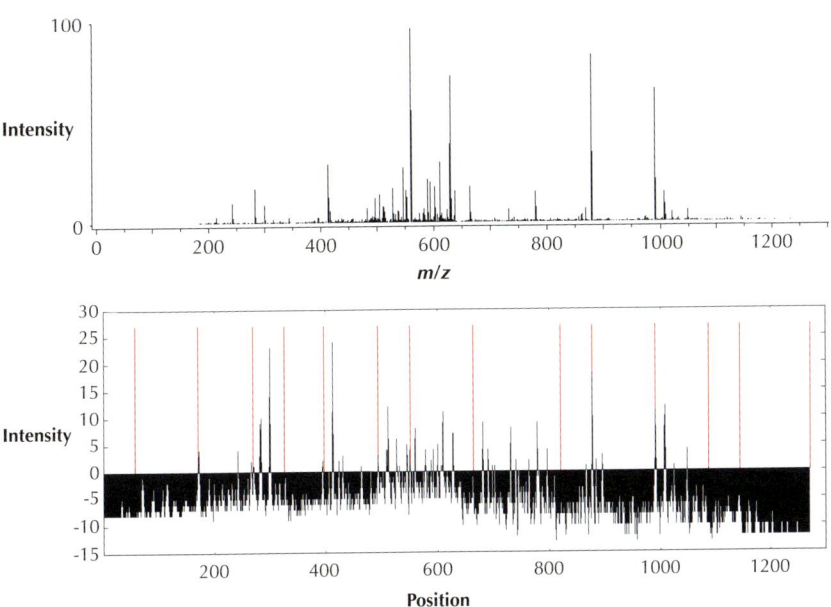

FIGURE 11.8 (Top) *DinosaurSpectrum* reproduced from Figure 11.5. (Bottom) The spectral vector of *DinosaurSpectrum*. The positions of ones in the peptide vector of GLVGAPGLRGLPGK (at coordinates 57, 170, 269, 326, 397, 494, 551, 664, 820, 877, 990, 1087, 1144, and 1272) are shown by red lines. The amplitudes of the spectral vector at these coordinates are equal to -8, +1, -4, -6, -6, +3, +1, -4, -8, +18, +11, -10, -7, and 0, respectively. SCORE(GLVGAPGLRGLPGK, *DinosaurSpectrum*) is the dot product of the peptide vector and spectral vector, $-8 + 1 - 4 - 6 - 6 + 3 + 1 - 4 - 8 + 18 + 11 - 10 - 7 + 0 = -19$. Because most amplitudes are negative, the fact that SCORE(GLVGAPGLRGLPGK, *DinosaurSpectrum*) is negative does not necessarily imply that the spectral interpretation above is incorrect.

STOP and Think: Can you find a peptide vector that scores higher than 22 against the spectral vector in Figure 11.7?

In the remainder of this chapter, we will work with spectral vectors instead of spectra. Given a spectral vector $\overrightarrow{Spectrum}$, our goal is to find a peptide *Peptide* maximizing

SCORE($\overrightarrow{Peptide}, \overrightarrow{Spectrum}$). Since the mass of a peptide and the parent mass of the spectrum that it generates should be the same, a peptide vector should have the same length as the spectral vector under consideration. We will therefore define the score between a peptide vector and a spectral vector of different length as $-\infty$.

Peptide Sequencing Problem:

Given a spectral vector, find a peptide with maximum score against this spectrum.

Input: A spectral vector $\overrightarrow{Spectrum}$.

Output: An amino acid string *Peptide* that maximizes SCORE($\overrightarrow{Peptide}$, $\overrightarrow{Spectrum}$) among all possible amino acid strings.

Peptide sequencing algorithm

Given a spectral vector $\overrightarrow{Spectrum} = (s_1, \ldots, s_m)$, we will construct a DAG on $m + 1$ nodes, labeled with the integers from 0 (*source*) to m (*sink*), and then connect node i to node j by a directed edge if $j - i$ is equal to the mass of an amino acid (Figure 11.9). We will further assign weight s_i to node i (for $1 \leq i \leq m$) and assign weight zero to node 0.

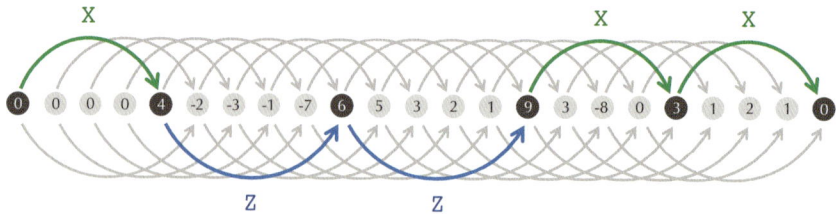

FIGURE 11.9 The node-weighted DAG for a spectral vector of length $m = 22$ and the amino acid alphabet {**X**, **Z**} with respective masses 4 and 5. The path from 0 to m representing peptide **XZZXX** (corresponding to the peptide vector (0, 0, 0, **1**, 0, 0, 0, 0, **1**, 0, 0, 0, 0, **1**, 0, 0, 0, **1**, 0, 0, 0, **1**) with score 0 + 4 + 6 + 9 + 3 + 0 is highlighted.

 **STOP and Think:** How does this DAG compare to the DAG GRAPH(*Spectrum*) that we constructed to decode an ideal spectrum?

Any path connecting *source* to *sink* in this DAG corresponds to an amino acid string *Peptide*, and the total weight of nodes on this path is equal to SCORE($\overrightarrow{Peptide}, \overrightarrow{Spectrum}$).

We have therefore reduced the Peptide Sequencing Problem to the problem of finding a maximum-weight path from *source* to *sink* in a node-weighted DAG.

STOP and Think: In Chapter 5, we developed an algorithm for finding a path of maximum weight in an *edge-weighted* DAG. How can we modify this algorithm to find a path of maximum weight in a *node-weighted* DAG?

EXERCISE BREAK: Apply your algorithm for the Peptide Sequencing Problem to $\overrightarrow{DinosaurSpectrum}$.

By applying an algorithm solving the Peptide Sequencing Problem to $\overrightarrow{DinosaurSpectrum}$, we find the peptide `ATKIVDCFMTY` with score 96 (Figure 11.5 (middle)). However, Asara proposed a different peptide, `GLVGAPGLRGLPGK` with score -19, which we will call *DinosaurPeptide* (Figure 11.5 (bottom)). This peptide has much lower score than `ATKIVDCFMTY`; in fact, *billions* of peptides outscore *DinosaurPeptide*!

STOP and Think: Why do you think that Asara proposed *DinosaurPeptide* instead of the higher-scoring `ATKIVDCFMTY`?

Peptide Identification

The Peptide Identification Problem

If you followed our struggles to sequence antibiotic peptides, then you will agree that we should be wary of jumping to the conclusion that the highest-scoring peptide for *DinosaurSpectrum* must have generated this spectrum.

Despite many attempts, researchers have still not devised a scoring function that reliably assigns the highest score to the biologically correct peptide, i.e., the peptide that generated the spectrum. Fortunately, although the correct peptide often does not achieve the highest score among *all* peptides, it typically does score highest among all peptides *limited* to the species's proteome. As a result, we can transition from peptide sequencing to peptide identification by limiting our search to peptides present in the proteome, which we concatenate into a single amino acid string *Proteome*.

CHAPTER 11

EXERCISE BREAK: How does the number of all peptides of length 10 (which we must explore in peptide sequencing) compare with the number of peptides of length 10 in the human proteome? (Note: there are approximately 20,000 protein-coding genes in the human protein, and the average length of a human protein is approximately 400 amino acids.)

Peptide Identification Problem:
Find a peptide from a proteome with maximum score against a spectrum.

Input: A spectral vector $\overrightarrow{Spectrum}$ and an amino acid string *Proteome*.
Output: An amino acid string *Peptide* that maximizes SCORE($\overrightarrow{Peptide}$, $\overrightarrow{Spectrum}$) among all substrings of *Proteome*.

STOP and Think: In practice, the input to the Peptide Identification Problem is a set of proteins rather than a single string *Proteome*. What are the potential pitfalls of concatenating all proteins as opposed to analyzing each protein separately?

Identifying peptides in the unknown T. rex proteome

You may be wondering why we have returned to peptide identification, since we do not know the *T. rex* proteome. It may therefore seem that we cannot apply an algorithm for the Peptide Identification Problem to the spectra obtained from the *T. rex* fossil.

STOP and Think: How could we form a protein database to search for *T. rex* peptides?

Approximately 90% of proteins making up animal bones are **collagen**. Dinosaur bones undoubtedly contained collagen, and there is little chance that other proteins could have survived for millions of years. Since the amino acid sequences of collagens are often conserved across different species, Asara reasoned that any proteins that had survived in the *T. rex* fossil would likely be similar to collagens from present-day species.

As a sanity check, Asara compared the *T. rex* spectra against the entire **UniProt** database, containing proteins from hundreds of species and totaling almost 200 million amino acids. He also included some mutated versions of collagens from present-day species in order to model possible differences between these collagens and collagens in *T. rex* (we will call the resulting protein database UniProt+). It turned out that most of the high-scoring peptides identified in this database were chicken collagens, which supported the hypothesis that birds evolved from dinosaurs.

In fact, *DinosaurPeptide* is only one mutation apart from a chicken collagen peptide. But how can we test whether this peptide is the correct interpretation of *DinosaurSpectrum*?

Searching for peptide-spectrum matches

Like peptide sequencing algorithms, peptide identification algorithms may return an erroneous peptide, particularly if the score of the highest-scoring peptide found in the proteome is much lower than the score of the highest-scoring peptide over all peptides. For this reason, biologists usually establish a score threshold and only pay attention to a solution of the Peptide Identification Problem if its score is at least equal to the threshold.

Given a set of spectral vectors *SpectralVectors*, an amino acid string *Proteome*, and a score threshold *threshold*, we will solve the Peptide Identification Problem for each vector $\overrightarrow{Spectrum}$ in *SpectralVectors* and identify a peptide *Peptide* having maximum score for this spectral vector over all peptides in *Proteome* (ties are broken arbitrarily). If SCORE($\overrightarrow{Peptide}, \overrightarrow{Spectrum}$) is greater than or equal to *threshold*, then we conclude that *Peptide* is present in the sample and call the pair (*Peptide*, $\overrightarrow{Spectrum}$) a **peptide-spectrum match (PSM)**. The resulting collection of PSMs for *SpectralVectors* is denoted PSM$_{threshold}$(*Proteome, SpectralVectors*).

PSM Search Problem:

Identify all peptide-spectrum matches scoring above a threshold for a set of spectra and a proteome.

> **Input**: A set of spectral vectors *SpectralVectors*, an amino acid string *Proteome*, and an integer *threshold*.
> **Output**: The set PSM$_{threshold}$(*Proteome, SpectralVectors*).

The following pseudocode solves the PSM Search Problem using an algorithm that you just implemented to solve the Peptide identification Problem, which we call PEPTIDEIDENTIFICATION.

Chapter 11

> **PSMSEARCH**(*SpectralVectors*, *Proteome*, *threshold*)
>   *PSMSet* ← an empty set
>   **for** each vector $\overrightarrow{Spectrum}$ in *SpectralVectors*
>   *Peptide* ← **PEPTIDEIDENTIFICATION**($\overrightarrow{Spectrum}$, *Proteome*)
>   **if** SCORE(*Peptide*, $\overrightarrow{Spectrum}$) ≥ *threshold*
>   add the PSM (*Peptide*, $\overrightarrow{Spectrum}$) to *PSMSet*
>   **return** *PSMSet*

DinosaurPeptide turned out to be the highest scoring peptide for *DinosaurSpectrum* among all peptides in the UniProt+ database. But do the billions of peptides not occurring in this database that outscore *DinosaurPeptide* imply that the database that Asara formed is incomplete and that *DinosaurSpectrum* arose from a different peptide?

The reality is that the highest scoring peptide in a proteome is commonly outscored by billions of peptides not belonging to the proteome. However, this phenomenon does not imply that **PSMSEARCH** has identified the wrong peptide, because the total number of peptides with the same mass may be measured in the trillions or even quadrillions. In other words, the billions of peptides that outscore *DinosaurPeptide* represent a small fraction of all peptides having the same mass as this peptide. Thus, we need to complement **PSMSEARCH** with an evaluation of the statistical significance of its identified PSMs.

> **STOP and Think:** Suppose that we search 1,000 spectra from a chicken sample against the chicken proteome, and identify 100 PSMs whose score surpasses a threshold. How would you estimate the percentage of erroneous PSMs among these 100 PSMs?

Peptide Identification and the Infinite Monkey Theorem

False discovery rate

To estimate the number of spurious PSMs in $\text{PSM}_{threshold}$(*Proteome*, *SpectralVectors*), we will construct a **decoy proteome** *DecoyProteome*, a randomly generated amino acid string having the same length as *Proteome* (with the probability of generating any amino acid at each position equal to 1/20). We will then solve the PSM Search Problem for *DecoyProteome* instead of *Proteome* for the same score threshold.

We are not interested in the PSMs identified in the randomly generated decoy proteome, which are nothing more than biologically irrelevant artifacts. The point

WAS T. REX JUST A BIG CHICKEN?

is that the *number* of these PSMs will provide a rough estimate for the number of erroneous PSMs identified in our biologically relevant search against the real proteome. We will therefore define the **false discovery rate (FDR)** of a PSM search as the ratio of the number of decoy PSMs to the number of PSMs identified with respect to the real proteome,

$$\frac{|\text{PSM}_{threshold}(DecoyProteome, SpectralVectors)|}{|\text{PSM}_{threshold}(Proteome, SpectralVectors)|}.$$

For example, if a search against *Proteome* results in 100 PSMs, and a search against *DecoyProteome* results in just five PSMs, then the FDR would be 5%, and we would conclude that approximately 95% of identified PSMs are likely valid. On the other hand, if searching against *DecoyProteome* returned close to 100 PSMs, then the FDR would be close to 1, and we would have a hard time making the argument that any peptides identified in our search against *Proteome* are biologically relevant.

EXERCISE BREAK: Estimate the FDR when searching all *T. rex* spectra against the UniProt+ database with *threshold* = 80.

STOP and Think: If the FDR turns out to be very high for a given value of *threshold*, can we still find reliable PSMs?

Even if the FDR is high, then we should not conclude that our spectral dataset is worthless or that we are searching against the wrong protein database. We may simply have selected the wrong score threshold for analyzing our data, since the FDR can vary widely depending on the choice of this threshold (Figure 11.10).

For the *T. rex* spectral dataset, we find 27 PSMs in the amino acid string *Proteome* derived from UniProt+ and only one PSM in *DecoyProteome* with score at least equal to *threshold* = 100 (FDR = 3.7%). Unfortunately, we cannot automatically conclude that we have discovered two dozen dinosaur peptides because many of these PSMs correspond to common laboratory contaminants. Our goal is to figure out whether the remaining few, including *DinosaurPeptide*, indeed correspond to dinosaur peptides.

Specifically, FDR helps us analyze the *entire* set of identified PSMs for all *T. rex* spectra, but what about the statistical significance of an *individual* PSM? In particular, can we determine whether the PSM (*DinosaurPeptide*, $\overrightarrow{DinosaurSpectrum}$), which we will call *DinosaurPSM*, is statistically significant? To answer this question, we first need to quantify what we mean by "statistically significant".

CHAPTER 11

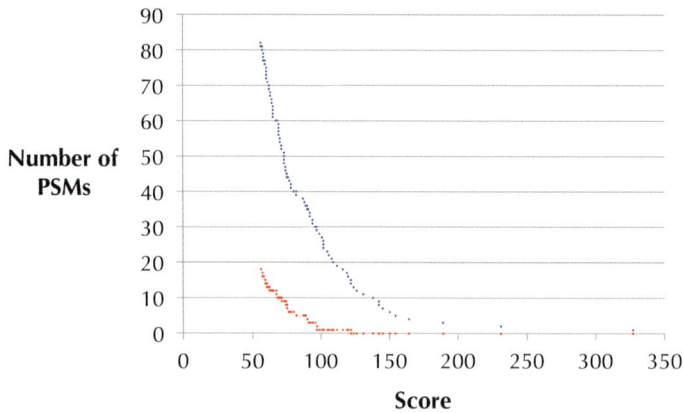

FIGURE 11.10 The number of PSMs identified by searching *DinosaurSpectrum* against the amino acid string derived from the UniProt+ database (blue) and a decoy proteome of the same length (red) depending on the score threshold.

The monkey and the typewriter

Imagine that we have locked you in a room with a monkey and a typewriter. The monkey quickly tires of your company and starts banging away on the typewriter, creating strings of symbols (we will assume that the monkey particularly likes the space bar). As you begin to lose your sanity, you test each new string generated by the monkey to see if some of them are correctly spelled English words. (Figure 11.11). After all, according to the **infinite monkey theorem**, the monkey will eventually type *Hamlet* (see **DETOUR: The Infinite Monkey Theorem**).

FIGURE 11.11 Searching for William Shakespeare.

You would be shocked if the monkey immediately typed "To be, or not to be". But would you be as surprised if, after cranking out a million strings, the monkey had typed a dozen English words?

Given a set of strings *Dictionary*, we define E(*Dictionary*, *n*) as the expected number of occurrences of strings from *Dictionary* in a randomly generated string of length *n*, where the probabilities of generating every letter at each position of the string are the same. Let *EnglishDictionary* denote the set of all English words. If it turns out that, after typing *n* symbols, the monkey types significantly more English words than E(*EnglishDictionary*, *n*), then we have every reason to believe that the monkey can spell! On the other hand, if the monkey types about the same number of words as E(*EnglishDictionary*, *n*), then the monkey is probably not Shakespeare reincarnate.

The Monkey and the Typewriter Problem:

Find the expected number of strings from a dictionary appearing in a randomly generated text.

Input: A set of strings *Dictionary* and an integer *n*.
Output: E(*Dictionary*, *n*).

STOP and Think: What do the monkey and the typewriter have to do with mass spectrometry?

EXERCISE BREAK: What is the expected number of times that the string SHAKESPEARE appears in a randomly generated English string (with no spaces) of length 200 million?

Statistical significance of a peptide-spectrum match

Now imagine that instead of a monkey typing words, we have an algorithm generating the set of all peptides scoring at least *threshold* against a spectral vector $\overrightarrow{Spectrum}$. We will henceforth call this set of high-scoring peptides a **spectral dictionary**, denoted

$$\text{DICTIONARY}_{threshold}(\overrightarrow{Spectrum}).$$

For a PSM (*Peptide*, $\overrightarrow{Spectrum}$), we will use the term **PSM dictionary**, denoted DICTIONARY(*Peptide*, $\overrightarrow{Spectrum}$), to refer to the spectral dictionary

CHAPTER 11

$$\text{DICTIONARY}_{\text{SCORE}(Peptide, \overrightarrow{Spectrum})}(\overrightarrow{Spectrum}).$$

For *DinosaurPSM*, the PSM dictionary is $\text{DICTIONARY}_{-19}(\overrightarrow{DinosaurSpectrum})$.

Instead of checking which words generated by the monkey occur in an English dictionary, we will match peptides from a spectral dictionary against a proteome. If we find matches, then we must decide whether these matches represent biologically valid PSMs or just statistical artifacts. To make this decision, we consider

$$E(\text{DICTIONARY}_{threshold}(\overrightarrow{Spectrum}), n),$$

the expected number of peptides in a decoy proteome of length n that would occur in $\text{DICTIONARY}_{threshold}(\overrightarrow{Spectrum})$. If this number is larger than 1, then there is nothing surprising in finding a peptide that scores *threshold* against $\overrightarrow{Spectrum}$. We have therefore formulated our statistical significance test as a special case of the Monkey and the Typewriter Problem.

Expected Number of High-Scoring Peptides Problem:

Find the expected number of high-scoring peptides against a given spectrum in a decoy proteome.

Input: A spectral vector $\overrightarrow{Spectrum}$ and integers *threshold* and n.
Output: $E(\text{DICTIONARY}_{threshold}(\overrightarrow{Spectrum}), n)$.

To solve this problem, we will begin with a spectral dictionary consisting of a single amino acid string *Peptide*, which we will attempt to match against a randomly generated string *DecoyProteome* of length n. Because *DecoyProteome* was randomly generated, the probability that *Peptide* matches the string beginning at a given position of *DecoyProteome* is $1/20^{|Peptide|}$. We call this expression the **probability** of *Peptide*. Therefore, the expected number of times that *Peptide* occurs in *DecoyProteome* is

$$\frac{n - |Peptide| + 1}{20^{|Peptide|}} \approx n \cdot \frac{1}{20^{|Peptide|}}.$$

Next, assume that a set of peptides *Dictionary* contains multiple amino acid strings of arbitrary lengths. Using the above approximation, the expected number of matches between strings in *Dictionary* and *DecoyProteome* can be approximated as

$$E(Dictionary, n) \approx n \cdot \left(\sum_{\text{each peptide } Peptide \text{ in } Dictionary} \frac{1}{20^{|Peptide|}} \right).$$

We will refer to the sum inside the parentheses above as the **probability** of *Dictionary*, denoted Pr(*Dictionary*), so that the preceding approximation can be written as

$$E(Dictionary, n) \approx n \cdot \Pr(Dictionary).$$

We have thus reduced the statistical analysis of a PSM (*Peptide*, $\overrightarrow{Spectrum}$) to the computation of Pr(DICTIONARY(*Peptide*, $\overrightarrow{Spectrum}$)), the probability of the PSM dictionary.

STOP and Think: Can Pr(DICTIONARY(*Peptide*, $\overrightarrow{Spectrum}$)) be greater than 1?

You may be wondering why we have used the probabilistic notation Pr(*Dictionary*). To learn why Pr(DICTIONARY(*Peptide*, $\overrightarrow{Spectrum}$)) is indeed a probability (and thus cannot exceed 1), see **DETOUR: The Probabilistic Space of Peptides in a Spectral Dictionary.**

Probability of Spectral Dictionary Problem:
Find the probability of a spectral dictionary for a given spectrum and score threshold.

Input: A spectral vector $\overrightarrow{Spectrum}$ and an integer *threshold*.
Output: The probability of DICTIONARY$_{threshold}$($\overrightarrow{Spectrum}$).

It seems that we are finally ready to test the statistical significance of *DinosaurPSM*. We simply need to first construct the PSM dictionary DICTIONARY(*DinosaurPSM*) and then compute $n \cdot$ Pr(DICTIONARY(*DinosaurPSM*)), where n is the length of the string formed by concatenating the UniProt+ database. If this value is small (say, 0.001), then we will be able to argue that *DinosaurPeptide* is a *T. rex* peptide rather than a statistical artifact.

Unfortunately, DICTIONARY(*DinosaurPSM*) contains over 200 *billion* peptides, and generating it would be extremely time-consuming. Can we somehow compute the probability of this dictionary without having to generate it?

CHAPTER 11

Spectral Dictionaries

We will first compute the number of peptides in a spectral dictionary, since this simpler problem will provide insights on how to compute the probability of a spectral dictionary.

Size of Spectral Dictionary Problem:

Find the size of the spectral dictionary for a given spectral vector and score threshold.

Input: A spectral vector $\overrightarrow{Spectrum}$ and an integer *threshold*.
Output: The number of peptides in $\text{DICTIONARY}_{threshold}(\overrightarrow{Spectrum})$.

We will use dynamic programming to solve the Size of Spectral Dictionary Problem. Given a spectral vector $\overrightarrow{Spectrum} = (s_1, \ldots, s_m)$, we define its *i*-**prefix** (for *i* between 1 and *m*) as $\overrightarrow{Spectrum}_i = (s_1, \ldots, s_i)$ and introduce a variable $\text{SIZE}(i, t)$ as the number of peptides *Peptide* of mass *i* such that $\text{SCORE}(Peptide, \overrightarrow{Spectrum}_i)$ is equal to *t*. For example, consider the spectral vector $\overrightarrow{Spectrum} = (4, -3, -2, 3, 3, -4, 5, -3, -1, -1, 3, 4, 1, 3)$ of length 14 and the toy amino acid alphabet consisting of amino acids X and Z with respective masses 4 and 5. There are only three peptides of mass 13 (XXZ, XZX, and ZXX); the first two peptides have score 1 against $\overrightarrow{Spectrum}_{13}$, and the third has score 3. Thus, $\text{SIZE}(13, 1) = 2$, $\text{SIZE}(13, 3) = 1$, and $\text{SIZE}(13, t) = 0$ for all values of *t* other than 1 and 3.

The key to establishing a recurrence relation for computing $\text{SIZE}(i, t)$ is to realize that the set of peptides contributing to $\text{SIZE}(i, t)$ can be split into 20 subsets depending on their final amino acid *a*. Each peptide ending in a specific amino acid *a* results in a shorter peptide with mass $i - |a|$ and score $t - s_i$ if we remove *a* from the peptide (here, $|a|$ denotes the mass of *a*). Thus, as illustrated in Figure 11.12,

$$\text{SIZE}(i, t) = \sum_{\text{all amino acids } a} \text{SIZE}(i - |a|, t - s_i).$$

Since there is a single "empty" peptide of length zero, we initialize $\text{SIZE}(0, 0) = 1$. We also define $\text{SIZE}(0, t) = 0$ for all possible scores *t*, and set $\text{SIZE}(i, t) = 0$ for negative values of *i*. Using the above recurrence, we can compute the size of a spectral dictionary of $\overrightarrow{Spectrum} = (s_1, \ldots, s_m)$ as

$$\left| \text{DICTIONARY}_{threshold}(\overrightarrow{Spectrum}) \right| = \sum_{t \geq threshold} \text{SIZE}(m, t).$$

258

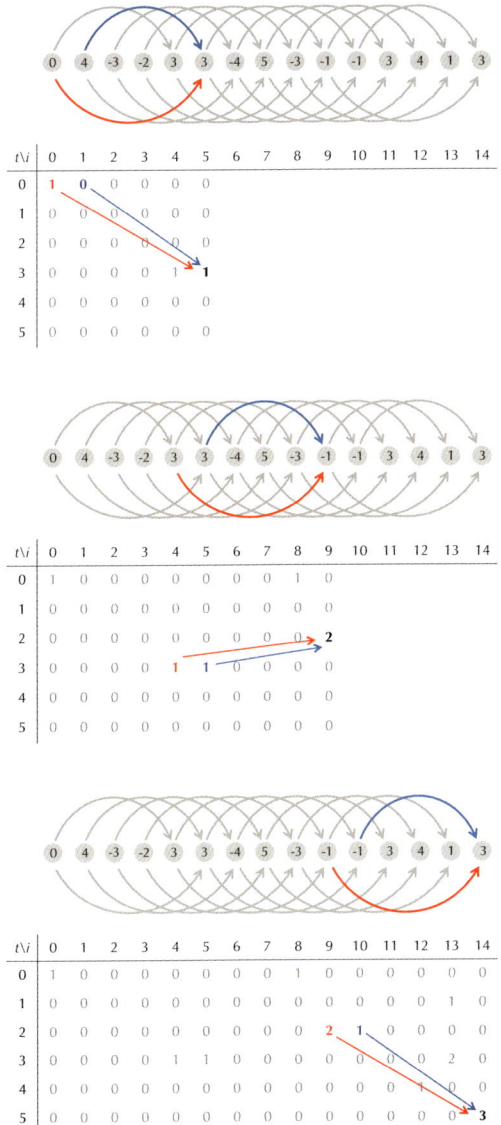

FIGURE 11.12 Computing SIZE(i, t) for the toy alphabet consisting of the amino acids X and Z with respective masses 4 and 5 and the spectral vector (4, -3, -2, 3, 3, -4, 5, -3, -1, -1, 3, 4, 1, 3). The boldfaced black entries in the dynamic programming matrices are computed by summing blue and red entries according to the formula SIZE(i, t) = SIZE$(i - 4, t - s_i)$ + SIZE$(i - 5, t - s_i)$. For example, in the matrix at the bottom, SIZE$(14, 5)$ = SIZE$(14 - 4, 5 - 3)$ + SIZE$(14 - 5, 5 - 3)$ = SIZE$(10, 2)$ + SIZE$(9, 2)$.

CHAPTER 11

STOP and Think: The above formula requires computing SIZE(m,t) for all values of t greater than threshold. Given a spectral vector $\overrightarrow{Spectrum}$, can you find a value T such that SIZE(m,t) is equal to zero when $t > T$?

EXERCISE BREAK: Compute |DICTIONARY($DinosaurPSM$)|.

Note that the equation for the *probability* of a dictionary,

$$\Pr(Dictionary) = \sum_{\text{each peptide } Peptide \text{ in } Dictionary} \frac{1}{20^{|Peptide|}},$$

is similar to an equation for the *size* of a dictionary,

$$|Dictionary| = \sum_{\text{each peptide } Peptide \text{ in } Dictionary} 1.$$

This similarity suggests that we can derive a recurrence for the probability of a dictionary using arguments similar to those used to find the size of a dictionary.

Define $\Pr(i,t)$ as the sum of probabilities of all peptides with mass i for which SCORE$(\overrightarrow{Peptide}, \overrightarrow{Spectrum_i})$ is equal to t. The set of peptides contributing to $\Pr(i,t)$ can be split into 20 subsets depending on their final amino acid. Each peptide $Peptide$ ending in a specific amino acid a results in a shorter peptide $Peptide_a$ if we remove a; $Peptide_a$ has mass $i - |a|$ and score $t - s_i$. Since the probability of $Peptide$ is 20 times smaller than the probability of $Peptide_a$, the contribution of $Peptide$ to $\Pr(i,t)$ is 20 times smaller than contribution of $Peptide_a$ to $\Pr(i - |a|, t - s_i)$. Therefore, $\Pr(i,t)$ can be computed as

$$\Pr(i,t) = \sum_{\text{all amino acids } a} \frac{1}{20} \cdot \Pr(i - |a|, t - s_i),$$

which differs from the recurrence for computing SIZE(i,t) only in the presence of the factor $1/20$.

We can now compute the probability of a spectral dictionary as

$$\Pr(\text{DICTIONARY}_{threshold}(\overrightarrow{Spectrum})) = \sum_{t \geq threshold} \Pr(m,t).$$

In particular, we find that DICTIONARY($DinosaurPSM$) consists of 219,136,251,374 peptides and has probability 0.00018. We are therefore ready to test the statistical significance of $DinosaurPSM$ found in searches against the UniProt+ database of length $n = 194,613,142$ (comprising 546,799 proteins).

Our aim is to compute $n \cdot \Pr(\text{DICTIONARY}(DinosaurPSM))$, as it approximates the number of peptides from $\text{DICTIONARY}(DinosaurPSM)$ that we expect to find in a decoy proteome of length n. Since $\Pr(\text{DICTIONARY}(DinosaurPSM)) = 0.00018$, we have that

$$n \cdot \Pr(\text{DICTIONARY}(DinosaurPSM)) = 35{,}311.$$

We therefore expect to find tens of thousands of peptides scoring at least as high as $DinosaurPeptide$ (against $\overrightarrow{DinosaurSpectrum}$) in a decoy database, and so there is nothing surprising about finding $DinosaurPSM$ while searching the UniProt+ database! We therefore conclude that $DinosaurPeptide$ is a statistical artifact rather than a real T. rex peptide. But what about the other T. rex peptides?

EXERCISE BREAK: Compute probabilities of spectral dictionaries for all other T. rex PSMs reported by Asara. Are these PSMs statistically significant?

T. rex Peptides: Contaminants or Treasure Trove of Ancient Proteins?

The hemoglobin riddle

Upon receiving criticism regarding the statistical foundations of his claims, Asara acknowledged some of the problems with his analysis, withdrew $DinosaurPeptide$ as an explanation for $DinosaurSpectrum$, changed some of his previously proposed T. rex peptides, and released all 31,372 spectra from the T. rex fossil. Afterwards, other scientists re-analyzed all spectra and verified that although some of the originally reported T. rex PSMs are questionable, others are statistically solid (Figure 11.13).

However, Asara's release of T. rex spectra raised more questions than it answered. In these spectra, Matthew Fitzgibbon and Martin McIntosh identified an additional spectrum (Figure 11.14) that perfectly matched ostrich hemoglobin, thus adding another T. rex peptide to the seven collagen peptides in Figure 11.13. The hemoglobin PSM, which was missed by Asara, is an order of magnitude more statistically significant than any previously reported T. rex collagen peptide!

It would be shocking if the hemoglobin peptide indeed belonged to T. rex because hemoglobins are much less conserved than collagens. For example, human beta chain hemoglobin is 146 amino acids long and has 27, 38, and 45 amino acid differences with mouse, kangaroo, and, chicken respectively. Furthermore, intact hemoglobin peptides have never been found in much younger and widely available fossils, such as the bones of extinct cave bears. These fossils are so common in European caves that they were used as a source of phosphates to produce gunpowder during World War I.

ID	Peptide	Protein	Probability	$n \cdot$ Probability
P1	GL**V**GAPGLRGLPGK	Collagen α1t2	$1.8 \cdot 10^{-4}$	36,000
P2	GVVGLP$_{oh}$GQR	Collagen α1t1	$7.6 \cdot 10^{-8}$	16
P3	GVQGPP$_{oh}$GPQGPR	Collagen α1t1	$7.9 \cdot 10^{-11}$	$1.6 \cdot 10^{-2}$
P4	GATGAP$_{oh}$GIAGAP$_{oh}$GFP$_{oh}$GAR	Collagen α1t1	$3.2 \cdot 10^{-12}$	$6.4 \cdot 10^{-4}$
P5	GLPGESGAVGPAGPIGSR	Collagen α2t1	$9.9 \cdot 10^{-14}$	$2.0 \cdot 10^{-5}$
P6	GSAGPP$_{oh}$GATGFP$_{oh}$GAAGR	Collagen α1t1	$3.2 \cdot 10^{-14}$	$6.4 \cdot 10^{-6}$
P7	GAPGPQGPSGAP$_{oh}$GP**K**	Collagen α1t1	$7.0 \cdot 10^{-16}$	$1.4 \cdot 10^{-7}$
P8	VNVADCGAEALAR	Hemoglobin β	$7.8 \cdot 10^{-17}$	$1.6 \cdot 10^{-8}$

FIGURE 11.13 The seven candidate *T. rex* collagen peptides (P1 - P7) reported by Asara as well as a hemoglobin peptide (P8). The last column shows the probabilities of the PSM dictionaries formed by these peptides. Red symbols indicate mutated amino acids compared to peptides in the UniProt database. The amino acid P_{oh} stands for hydroxyproline, a modified form of proline that is common in collagens.

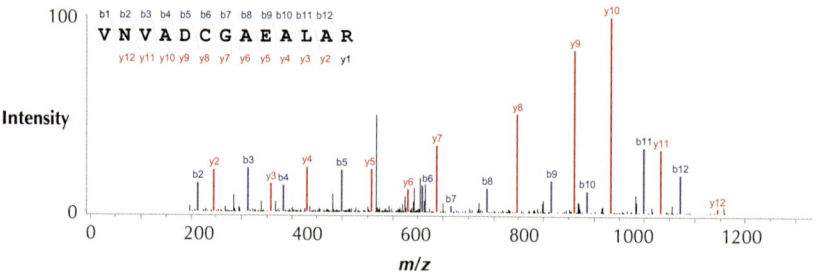

FIGURE 11.14 A high-quality *T. rex* spectrum matching the ostrich hemoglobin peptide VNVADCGAEAIAR. Nearly all possible prefixes and suffixes are represented by high-intensity peaks; in fact, applying *de novo* sequencing to this spectrum results in the same peptide.

Because Asara had analyzed ostrich samples before analyzing the *T. rex* sample, Fitzgibbon and McIntosh argued that the hemoglobin peptide might indicate a contaminated sample in the form of a **carryover**, or the identification of leftover peptides hiding inside a mass spectrometer after a previous experiment. Contamination is a fact of life in every proteomics laboratory, as mass spectrometrists are never surprised when they identify human keratin in their samples: the air in any room typically contains millions of tiny human skin particles.

If the hemoglobin peptide is a carryover, then the entire *T. rex* sample has been contaminated, implying that all other *T. rex* peptides should be discarded. However, Asara maintained that there was no contamination in his experiment and that the ostrich hemoglobin must be a *T. rex* peptide, expanding the class of proteins that can survive for 68 million years beyond just collagens.

Yet if Horner's *T. rex* fossil is indeed a treasure trove of ancient proteins, and we believe that the hemoglobin peptide came from *T. rex*, then why should we limit our search to collagen peptides and their mutated variants? Why not search against all known proteins from all vertebrates? Of course, we should use criteria that are similar to ones that Asara used, such as allowing for up to one mutation. If we follow this criterion, then we should supplement Figure 11.13 with a surprisingly diverse set of peptides from ostrich, chicken, mouse, and human; some of these peptides are shown in Figure 11.15.

In light of these new peptides, Asara's claim about finding *molecular* evidence of the link between birds and dinosaurs becomes even weaker (for more on the debate surrounding the claim that birds evolved from dinosaurs, see **DETOUR: Are Terrestrial Dinosaurs Really the Ancestors of Birds?**). If we were to attempt to dismiss the peptides in Figure 11.15 as statistical artifacts, then we might have to throw out the *T. rex* peptides in Figure 11.13 as well.

PAGE 279

ID	Peptide	Protein	Probability	n · Probability
P9	EDCLSG**A**KPK	ATG7 (Chicken)	$3.2 \cdot 10^{-12}$	$6.4 \cdot 10^{-4}$
P10	ENAGEDPGLAR	DCD (Human)	$2.7 \cdot 10^{-12}$	$5.4 \cdot 10^{-4}$
P11	**E**GVDAGAAGDPER	TTL11 (Mouse)	$1.2 \cdot 10^{-12}$	$2.4 \cdot 10^{-4}$
P12	S**W**IHVALVTGGNK	CBR1 (Human)	$1.2 \cdot 10^{-12}$	$2.4 \cdot 10^{-4}$
P13	SSN**V**LSGSTLR	MAMD1 (Human)	$5.9 \cdot 10^{-13}$	$1.8 \cdot 10^{-4}$
P14	DEVTPA**Y**VVVAR	ASPM (Mouse)	$1.9 \cdot 10^{-13}$	$3.8 \cdot 10^{-5}$
P15	**R**NVADCGAEALAR	HBB (Ostrich)	$3.5 \cdot 10^{-15}$	$7.0 \cdot 10^{-7}$

FIGURE 11.15 Matching *T. rex* spectra against all vertebrate proteins in the UniProt database (allowing for up to 1 mutation) reveals a diverse set of peptides. Red symbols indicate mutated amino acids. Note the presence of another ostrich hemoglobin peptide (P15), which is slightly heavier (by 57 daltons) than the previously reported hemoglobin peptide in Figure 11.13 (P8). This change in mass may represent either a mutation of V into R (as shown above) or a modification of an amino acid.

CHAPTER 11

The dinosaur DNA controversy

As the "*T. rex* peptides" paper continues to age, there is no end in sight to its controversy. Yet it was not the first paper to report the retrieval of genetic material from dinosaurs. In 1994, Scott Woodward announced that he had sequenced DNA from an 80 million year-old dinosaur bone. The most vehement critic of his finding was — believe it or not — Mary Schweitzer, who proved that Woodward had only sequenced contaminated human DNA.

The moral is that although we often present scientific discoveries as clear and incontrovertible, the reality is that some of the interesting avenues of modern science often fall short of this ideal. In a sense, the academic battleground is part of the appeal of becoming a scientist in the first place. But we also cannot help but wonder if we would have a conclusive answer to whether Horner's fossil really contained dinosaur peptides if it had originally been shared with dozens of independent researchers, who would have undoubtedly unearthed the shocking appearance of hemoglobin in the *T. rex* samples. Fittingly, in their criticism of Woodward's "dinosaur DNA" paper, Schweitzer wrote, "real advance in [paleontology] will come only when it is demonstrated that those studies can be replicated in independent laboratories."

Epilogue: From Unmodified to Modified Peptides

Post-translational modifications

The **PSMSEARCH** algorithm can only identify a peptide if it occurs in a proteome without mutations. Yet some of the peptides in Figure 11.13 are mutated.

STOP and Think: How could we generalize **PSMSEARCH** to find mutated peptides?

To find a highest-scoring peptide with up to *k* mutations matching a spectral vector, we could generate all mutated variants of all peptides in the proteome, concatenate them into an amino acid string *MutatedProteome*, and then run **PSMSEARCH** on *MutatedProteome*. Unfortunately, the number of mutated peptides will be so large that it may render **PSMSEARCH** impractical, even if we allow at most one mutation per peptide.

STOP and Think: How many mutated peptides with at most *k* mutations are there for a given peptide of length *n*?

In addition to searching for mutated peptides, we will also need to search for **post-translational modifications**, which alter amino acids after a protein has been translated from RNA. In fact, most proteins are modified after translation, and hundreds of types of modifications have been discovered. For example, the enzymatic activity of many proteins is regulated by the addition or removal of a phosphate group at a specific amino acid (Figure 11.16). This process, called **phosphorylation**, is reversible; **protein kinases** add phosphate groups, whereas **protein phosphatases** remove them.

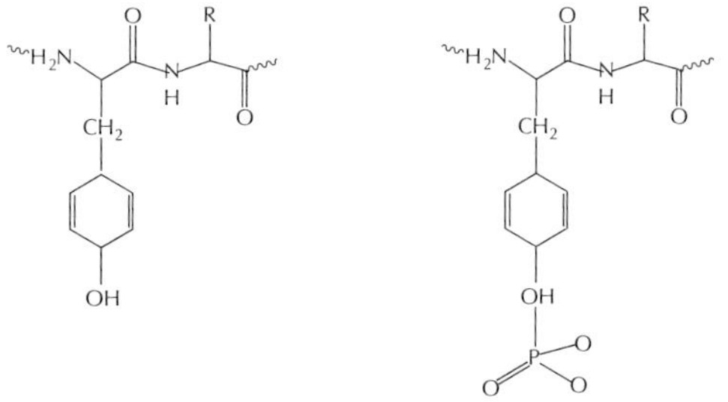

FIGURE 11.16 Tyrosine (left) and its post-translational modification into phosphorylated tyrosine (right).

In fact, you may have noticed in Figure 11.13 that most candidate *T. rex* peptides have a modification transforming proline (mass 97) into **hydroxyproline** (mass 113). Hydroxyproline is a major component of collagens that is important to collagen stability and that comprises roughly 4% of all amino acids in humans.

There are also important but rare post-translational modifications such as **diphthamide**. This modification of histidine only appears in a single protein (**protein synthesis elongation factor-2**), but it is universal across all eukaryotes! Researchers showed that diphthamide is the target for several toxins secreted by various pathogenic bacteria, which raises the question of why all eukaryotes would retain this modification if it makes them so vulnerable to pathogens — it must serve some important but still unknown function in normal physiology.

Searching for modifications as an alignment problem

A modification of mass δ applied to an amino acid results in adding δ to the mass of this amino acid. For example, $\delta = 80$ for phosphorylated amino acids (serine, threonine, and tyrosine), $\delta = 16$ for the modification of proline into hydroxyproline, and $\delta = -1$

for the modification of lysine into **allysin**. If δ is positive, then the resulting modified peptide has a peptide vector that differs from the original peptide vector $\overrightarrow{Peptide}$ by inserting a block of δ zeroes before the i-th occurrence of 1 in $\overrightarrow{Peptide}$. In the more rare case that δ is negative, the modified peptide corresponds to deleting a block of $|\delta|$ zeroes from $\overrightarrow{Peptide}$ (Figure 11.17).

We will use the term **block indel** to refer to the addition or removal of a block of consecutive zeroes from a binary vector. Thus, applying k modifications to an amino acid string $Peptide$ corresponds to applying k block indels to its peptide vector $\overrightarrow{Peptide}$. We define $\text{VARIANTS}_k(Peptide)$ as the set of all modified variants of $Peptide$ with up to k modifications.

Given a peptide $Peptide$ and a spectral vector $\overrightarrow{Spectrum}$, our goal is to find a modified peptide from $\text{VARIANTS}_k(Peptide)$ with maximum score against $\overrightarrow{Spectrum}$.

Spectral Alignment Problem:

Given a peptide and a spectral vector, find a modified variant of this peptide that maximizes the peptide-spectrum score, among all variants of the peptide with up to k modifications.

Input: An amino acid string $Peptide$, a spectral vector $\overrightarrow{Spectrum}$, and an integer k.
Output: A peptide of maximum score against $\overrightarrow{Spectrum}$ among all peptides in $\text{VARIANTS}_k(Peptide)$.

A brute force approach to the Spectral Alignment Problem would score each peptide in $\text{VARIANTS}_k(Peptide)$ against the spectrum. We need to solve this problem more efficiently because our more ambitious goal is to solve the following problem, which will require multiple applications of an algorithm solving the Spectral Alignment Problem.

Modification Search Problem:

Given a spectrum and a proteome, find a peptide of maximum score against this spectrum among all modified peptides in the proteome with up to k modifications.

Input: A spectral vector $\overrightarrow{Spectrum}$, an amino acid string $Proteome$ and an integer k.
Output: A peptide $Peptide$ that maximizes $\text{SCORE}(\overrightarrow{Peptide}, \overrightarrow{Spectrum})$ among all modified variants of peptides from $Proteome$ with up to k modifications.

Like modifications, mutations can also be viewed as block indels; for example, the mutation of V (integer mass 99) into R (integer mass 156) in peptide P9 from Figure 11.15 can be viewed as a block insertion with $\delta = 156 - 99 = 57$. We can therefore transform the Modification Search Problem into a "Mutation Search Problem" by simply substituting the word "modification" by "mutation" in the problem statement. In this new problem, a mutation of mass δ applied to an amino acid corresponds to the difference between the mass of this amino acid and another one. For example, mutations of valine (integer mass 99) correspond to modifications with the integer masses -42, -28, -12, -2, 2, 4, 14, 15, 16, 29, 30, 32, 38, 48, 57, 64, and 87.

```
XZZXX      0 0 0 1 0 0 0 0 1 0 0 0 0 1 0 0 0 1 0 0 0 1
XZ+3ZXX    0 0 0 1 0 0 0 0 0 0 0 1 0 0 0 0 1 0 0 0 1 0 0 0 1
XZ+3ZX-2X  0 0 0 1 0 0 0 0 0 0 0 1 0 0 0 0 1 0 1 0 0 0 1
```

FIGURE 11.17 Transforming the peptide XZZXX into the peptide XZ^{+3}ZXX corresponds to inserting a block of three zeroes (shown in red) before the second occurrence of 1 in the peptide vector of XZZXX. Transforming the peptide XZ^{+3}ZXX into the peptide XZ^{+3}ZX^{-2}X corresponds to deleting a block of two zeroes (shown in green) before the fourth occurrence of 1 in the peptide vector of XZ^{+3}ZXX.

Building a Manhattan grid for spectral alignment

The problem of finding a highest scoring peptide vector having up to k block indels recalls sequence alignment problems. This insight suggests that we should frame the Spectral Alignment Problem as an instance of the Longest Path in a DAG Problem.

STOP and Think: How would you build a DAG to solve this problem?

Consider the $(m+1) \times (m+\Delta+1)$ Manhattan grid in which every node (i, j) is connected to every node (i', j') for $0 \leq i < i' \leq m$ and $0 \leq j < j' \leq m + \Delta$ (Figure 11.18). We call this graph SOUTHEAST$(m, m + \Delta)$ and call the nodes $(0,0)$ and $(m, m + \Delta)$ *source* and *sink*, respectively.

Next, consider an amino acid string *Peptide* $= a_1 \ldots a_n$ of mass m and its modified variant *Peptide*$^{\text{mod}} = a'_1 \ldots a'_n$ of mass $m + \Delta$. Define the path PATH(*Peptide*, *Peptide*^{mod})

in SOUTHEAST$(m, m + \Delta)$ consisting of n edges:

$$(0,0) \to (\text{MASS}(a_1), \text{MASS}(a_1'))$$
$$\to (\text{MASS}(a_1 a_2), \text{MASS}(a_1' a_2'))$$
$$\to \ldots$$
$$\to (\text{MASS}(a_1 \ldots a_n), \text{MASS}(a_1' \ldots a_n'))$$
$$= (m, m + \Delta).$$

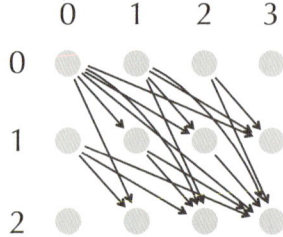

FIGURE 11.18 The graph SOUTHEAST(2, 3). Each node in the graph is connected to every node lying to its south and east, with the exception of nodes in the same row and column.

For example, consider the toy amino acid alphabet containing X, Y, and Z with respective masses 2, 3, and 4. The blue path in Figure 11.19 indicates PATH(XYYZX, **XY^{+2}YZ^{-3}X**):

$$(0,0) \to (2,2) \to (5,7) \to (8,10) \to (12,11) \to (14,13).$$

Except for the initial node (0, 0), every node (i, j) on this path indicates that the i-th element of $\overrightarrow{Peptide} = (0, 1, 0, 0, 1, 0, 0, 1, 0, 0, 0, 1, 0, 1)$ and the j-th element of $\overrightarrow{Peptide}^{\text{mod}}$ = (0, 1, 0, 0, 0, 0, 1, 0, 0, 1, 1, 0, 1) are both 1.

An edge connecting (i, j) to (i', j') in SOUTHEAST$(m, m + \Delta)$ is called **diagonal** if $i' - i = j' - j$ and **non-diagonal** otherwise. Note that if an amino acid of $Peptide^{\text{mod}}$ is unmodified, then the edge corresponding to this amino acid in PATH($Peptide, Peptide^{\text{mod}}$) is diagonal. An amino acid a with modification δ in $Peptide^{\text{mod}}$ corresponds to a non-diagonal edge in this path connecting some node (i, j) with the node $(i + |a|, j + |a| + \delta)$.

We are now ready to solve the Spectral Alignment Problem for an amino acid string $Peptide = a_1 \ldots a_n$ of mass m and a spectral vector $\overrightarrow{Spectrum} = s_1 \ldots s_{m+\Delta}$. We know that the modified variant of $Peptide$ solving this problem must have mass $m + \Delta$. We can thus represent all modified peptides of mass $m + \Delta$ in VARIANTS$_k$($Peptide$) as paths in SOUTHEAST$(m, m + \Delta)$ from *source* to *sink* with at most k non-diagonal edges (Figure 11.19 (top)). We refer to these modified peptides as VARIANTS$_k$(*Peptide*, $\overrightarrow{Spectrum}$).

STOP and Think: Does every path from *source* to *sink* in SOUTHEAST$(m, m + \Delta)$ correspond to a candidate modified peptide of mass $m + \Delta$?

Although every peptide in VARIANTS$_k$(*Peptide*, $\overrightarrow{Spectrum}$) corresponds to a path from *source* to *sink* in SOUTHEAST$(m, m + \Delta)$, many paths in this graph do not correspond to such modified peptides. Indeed, since *Peptide* is fixed, any path corresponding to a modified variant of *Peptide* will only pass through rows with indices

$$0, \text{MASS}(a_1), \text{MASS}(a_1 a_2), \ldots, \text{MASS}(a_1 \ldots a_n) = m,$$

shown as rows with darker nodes in Figure 11.19 (top).

Thus, nodes in other rows can be safely removed from SOUTHEAST$(m, m + \Delta)$, which results in the **PSM graph**, denoted PSMGRAPH(*Peptide*, $\overrightarrow{Spectrum}$), and shown in Figure 11.19 (bottom). Note that the $n + 1$ rows of nodes in the PSM graph have indices i equal to

$$0, \text{MASS}(a_1), \text{MASS}(a_1 a_2), \ldots, \text{MASS}(a_1 \ldots a_n)$$

rather than the indices $0, 1, \ldots, n$.

In the PSM graph, all edges entering into the row with index $i = \text{MASS}(a_1 \ldots a_t)$ originate at the row with index $\text{MASS}(a_1 \ldots a_{t-1})$. We therefore define DIFF(i) as the mass of the amino acid a_t. For the peptide XYYZX, DIFF$(2) = \text{MASS}(X) = 2$, DIFF$(5) = \text{MASS}(Y) = 3$, DIFF$(8) = \text{MASS}(Y) = 3$, DIFF$(12) = \text{MASS}(Z) = 4$, and DIFF$(14) = \text{MASS}(X) = 2$.

STOP and Think: Can you assign weights to the nodes of PSMGRAPH(*Peptide*, $\overrightarrow{Spectrum}$) so that the total weight of a path from *source* to *sink* corresponding to a modified peptide *Peptide*^{mod} is equal to SCORE(*Peptide*^{mod}, $\overrightarrow{Spectrum}$)?

Given a spectral vector $\overrightarrow{Spectrum} = (s_1, \ldots, s_{m+\Delta})$, we assign a weight of s_j to every node (i, j) in column j of the PSM graph. With this assignment of weights, the total weight of nodes in the path corresponding to *Peptide*^{mod} is equal to SCORE(*Peptide*^{mod}, $\overrightarrow{Spectrum}$). Solving the Spectral Alignment Problem is therefore equivalent to finding a path in the PSM graph with maximum total node weight among all paths with at most k non-diagonal edges.

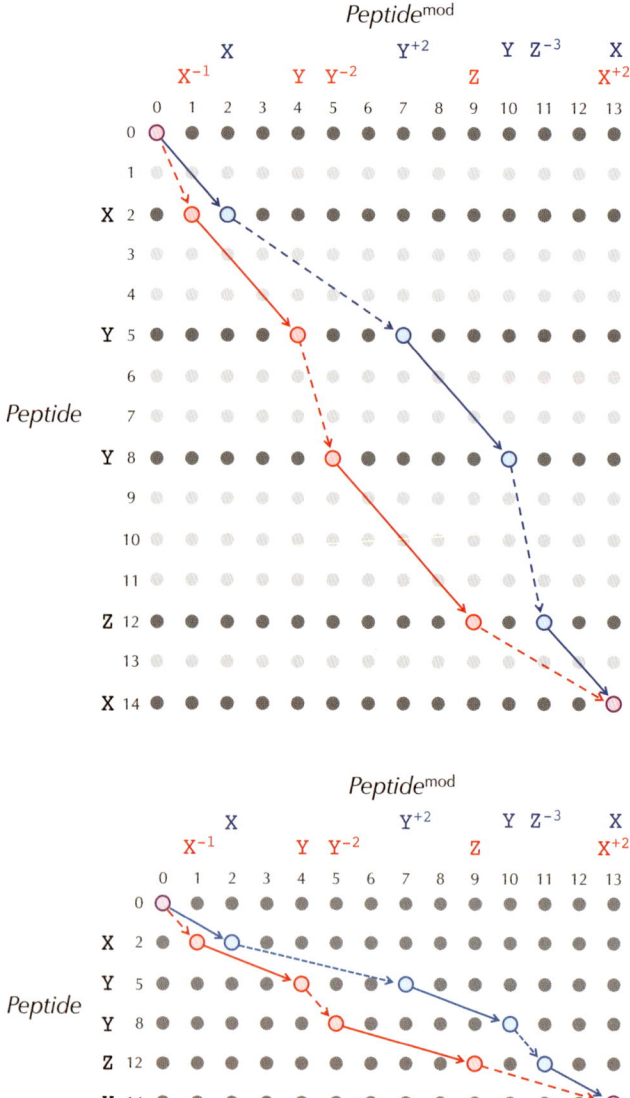

FIGURE 11.19 (Top) Two paths in SOUTHEAST(14, 13) formed by the modifications $\mathtt{XY^{+2}YZ^{-3}X}$ and $\mathtt{X^{-1}YY^{-2}ZX^{+2}}$ of XYYZX, where X, Y, and Z have respective masses 2, 3, and 4. Diagonal edges are solid, whereas non-diagonal edges are dashed. The indices of darker rows in the graph correspond to occurrences of 1 in the peptide vector of XYYZX. The darker nodes in this graph form the PSM graph. (Bottom) Another representation of the PSM graph obtained by removing the light nodes from the graph above.

Spectral alignment algorithm

We already know how to find a longest path in a node-weighted DAG. However, it is not clear how to find a longest path in a DAG under the additional constraint that this path has at most k non-diagonal edges.

As a workaround, we will convert the two-dimensional PSM graph from Figure 11.19 (bottom) into a three-dimensional **spectral alignment graph** consisting of $k+1$ layers (Figure 11.20), where the node set in each layer coincides with the node set of the PSM graph. This graph will have nodes (i, j, t), where $0 \le i \le m$, $0 \le j \le m + \Delta$, and $0 \le t \le k$. Each such node inherits the weight of the node (i, j) in the PSM graph (i.e., the amplitude s_j from $\overrightarrow{Spectrum} = (s_1, \ldots, s_{m+\Delta})$).

As for edges, each of the $k+1$ layers of the spectral alignment graph will inherit all diagonal edges from the PSM graph, i.e., each diagonal edge from (i, j) to $(i+x, j+x)$ in the PSM graph will correspond to the $k+1$ edges (i, j, t) to $(i+x, j+x, t)$ for $0 \le t \le k$.

STOP and Think: The layers in the constructed graph are now disconnected. How should we connect them?

For each non-diagonal edge connecting (i, j) to (i', j') in the PSM graph, we will generate k edges connecting consecutive layers in the spectral alignment graph by connecting (i, j, t) to $(i', j', t+1)$ for all t between 0 and $k-1$. Every path in the spectral alignment graph from $(0, 0, 0)$ to $(m, m + \Delta, t)$ corresponds to a modified version of peptide with t modifications (Figure 11.20). The zero-th layer of this graph will store scores of peptides with no modifications, the first layer will store scores of peptides with one modification, and so on.

To solve the Spectral Alignment Problem, we will define SCORE(i, j, t) as the maximum score of all paths connecting node $(0, 0, 0)$ to node (i, j, t) in the spectral alignment graph. Note that this score is equal to the weight s_j assigned to node (i, j, t) plus the maximum of the scores of all predecessors of node (i, j, t). One of these predecessors, $(i - \text{DIFF}(i), j - \text{DIFF}(i), t)$, is located in the same layer, and j other predecessors $((i - \text{DIFF}(i), j', t - 1)$ for $j' < j)$ are located in the previous layer. This reasoning results in the following recurrence for computing SCORE(i, j, t),

$$\text{SCORE}(i, j, t) = s_i + \max_{j' < j} \begin{cases} \text{SCORE}(i - \text{DIFF}(i), j - \text{DIFF}(i), t) \\ \text{SCORE}(i - \text{DIFF}(i), j', t - 1) \end{cases}.$$

The maximum score of all peptides with at most k modifications is therefore the maximum value of SCORE$(m, m + \Delta, t)$ as t ranges from 0 to k. To initialize the recurrence, we assume that SCORE$(0, 0, 0) = 0$ and that SCORE$(0, 0, t) = -\infty$ for all $1 \le t \le k$.

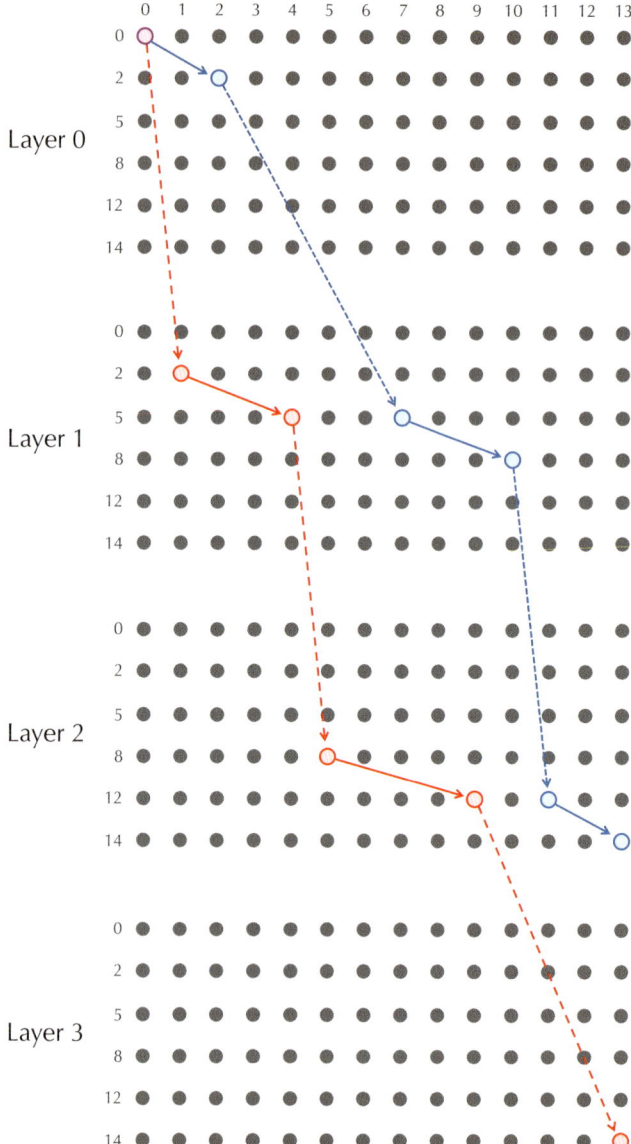

FIGURE 11.20 The PSM graph from Figure 11.19 (bottom) transformed into a spectral alignment graph with four layers ($k = 3$). The blue and red paths from the PSM graph in Figure 11.19 (bottom) correspond to the blue and red paths shown, which represent the respective modified variants $\mathtt{XY^{+2}YZ^{-3}X}$ and $\mathtt{X^{-1}YY^{-2}ZX^{+2}}$ of $\mathtt{XYYZX}$. The blue path terminates in Layer 2 because it corresponds to a peptide with two modifications, whereas the red path terminates in Layer 3 because it corresponds to a peptide with three modifications.

Although the above recurrence computes the score of a modified peptide solving the Spectral Alignment Problem, we also need to reconstruct this peptide. In order to achieve this goal, we will need to implement a backtracking approach similar to the backtracking approach described in Chapter 5, which we leave to you as an exercise.

EXERCISE BREAK: What is the running time of the spectral alignment algorithm?

CHALLENGE PROBLEM: In addition to *T. rex*, Asara also analyzed a 200,000 year-old mastodon fossil. The extinction of the elephant-like mastodons 10,000 years ago was caused by a combination of climate change and hunting by humans armed with stone weapons. In contrast to dinosaurs, it is not surprising that scientists routinely identify peptides from recently extinct species like mastodons or cave bears.

Analyze the mastodon collagen peptides reported in Asara's 2007 paper and decide which of them form statistically significant PSMs. Can you identify other statistically significant mastodon peptides that this paper missed? Can you find non-collagen peptides (especially hemoglobin peptides) matching spectra from mastodons? Can you determine the number of different types of post-translational modifications of mastodon peptides by solving the Modification Search Problem?

Detours

Gene prediction

To predict split genes, researchers often attempt to recognize the locations of splicing signals at exon-intron junctions. For a simple example, the dinucleotides `AG` and `GT` on either side of an exon are highly conserved (Figure 11.21). To improve the accuracy of this approach (known as **statistical gene prediction**), researchers look for genomic features appearing frequently in exons and infrequently in introns.

Attempts to improve the accuracy of statistical gene prediction methods have led to **similarity-based gene prediction** approaches, which are based on the observation that a newly sequenced gene is often similar to a known gene in another species. For example, 99% of mouse genes have human analogs.

However, we cannot simply look for a similar sequence in the mouse genome based on known human genes, since the exon sequence and partition of a gene into exons in different species may be different. To address this complication, similarity-based approaches sometimes look for a set of putative exons in the mouse genome whose concatenation fits a known human protein.

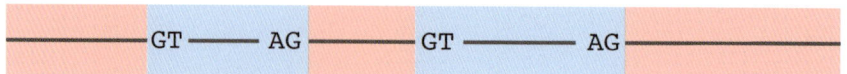

FIGURE 11.21 A split gene with exons (red) separated by introns (blue). Introns typically start with `GT` and end with `AG`.

Yet not all genes are split genes. In fact, bacteria do not have split genes at all, which simplifies bacterial gene prediction. Such genes begin with a start codon that codes for methionine (typically `ATG` but also sometimes `GTG` or `TTG`) and end with a stop codon (`TAA`, `TAG`, or `TGA`).

We can represent a genome of length n as a sequence of $n/3$ codons. The stop codons break this sequence into segments between every pair of consecutive stop codons. The suffixes of these segments that begin at the first start codon within a segment are called **open reading frames**, or **ORFs**. ORFs within a single genome may overlap because there are six possible reading frames.

In a randomly generated DNA string, we would expect to find a stop codon approximately every 64 nucleotides within a given reading frame. Yet the typical length of a bacterial gene is on the order of 1,000 nucleotides. A gene prediction algorithm might therefore select ORFs longer than some threshold length as candidate genes. Unfortunately, such an algorithm would also fail to detect short genes.

Many gene prediction algorithms also rely on subtle statistical differences between coding and non-coding regions, such as biases in **codon usage**, or the frequency of each codon. For example, there are six codons encoding leucine, but whereas `CUG` encodes 47% of all occurrences of leucine in *E. coli*, `CUA` encodes only 4%. Therefore, an ORF with many more occurrences of `CUG` than `CUA` is a candidate gene.

Bacterial gene prediction also takes advantage of several conserved motifs often found in genomic regions near start of RNA transcription. For example, the **Pribnow box** is a six-nucleotide sequence with consensus `TATAAT` that is an essential component for initiating transcription in bacteria.

Finding all paths in a graph

In the 19th Century, Charles Pierre Trémaux developed an algorithm for navigating mazes, described as follows. As you walk through a maze, drag a piece of chalk along the ground behind you. When you reach a junction, unmarked paths correspond to yet unexplored paths. So take an unexplored path for as long as you can, until you encounter a dead-end or a junction at which all the outgoing paths are marked. In this case, backtrack your steps until you encounter the exit or a junction with an unmarked path, or until you arrive at your starting point, in which case the maze has no solution.

Trémaux's maze algorithm is an example of **depth-first search (DFS)**, a technique for traversing the nodes of a graph. DFS starts at a given node and explores the graph as far as possible until it reaches a node that has no outgoing edges or from which we have already explored all edges. We then backtrack until we reach a node with unexplored edges. The preorder traversal that we encountered in DETOUR: From Suffix Trees to Suffix Arrays offers one example of a DFS applied to rooted trees.

The following recursive algorithm offers a DFS-inspired approach to finding all paths between a node v and a node *sink* in a DAG *Graph*.

ALLPATHS(*Graph, v, sink*)
    if v = *sink*
        Paths ← the set of paths consisting of the single-node path v
    else
        Paths ← an empty set of paths
        for all outgoing edges (v, w) from v
            PathsFromDescendant ← **ALLPATHS**(*Graph, v, sink*)
            add (v, w) as the first edge to each path in *PathsFromDescendant*
            add *PathsFromDescendant* to *Paths*
    return *Paths*

The Anti-Symmetric Path Problem

In the main text, we saw that not every path from *source* to *sink* in GRAPH(*Spectrum*) represents a solution of the Decoding an Ideal Spectrum Problem. This issue is caused by the fact that every mass in the spectrum may be interpreted as either a mass of a prefix or a mass of a suffix. Therefore, every node corresponding to mass s has a "twin" node (corresponding to MASS(*Peptide*) − s). Given an arbitrary node and its twin in GRAPH(*Spectrum*), a correct path from *source* to *sink* must pass through exactly one of these nodes in order to spell out a solution.

The Decoding an Ideal Spectrum Problem is a particular case of the following more general problem. Given a collection of **forbidden pairs** of nodes in a graph (when reconstructing peptides, forbidden pairs correspond to twins), a path in a graph is called **anti-symmetric** if it contains exactly one node from every forbidden pair.

Anti-Symmetric Path Problem:
Find an anti-symmetric path in a DAG.

> **Input**: A DAG with nodes *source* and *sink* and a set of forbidden pairs of nodes in this DAG.
> **Output**: An anti-symmetric path in this DAG from *source* to *sink*.

The Anti-Symmetric Path Problem is *NP*-Hard, but we should not give up hope of finding an efficient algorithm for the Decoding an Ideal Spectrum Problem because the latter is a specific instance of the former. In particular, forbidden pairs in peptide sequencing have the additional property that the sum of masses from each forbidden pair is equal to the mass of the entire peptide. In fact, there is a polynomial algorithm solving the Anti-Symmetric Path Problem for a DAG satisfying this additional property, but this algorithm is outside the scope of this book.

Transforming spectra into spectral vectors

Our goal is to develop a probabilistic model describing how a peptide vector generates an *integer-valued* spectrum and to use this model to transform a spectrum into a spectral vector. To address this problem, we will first introduce an abstract model that seemingly has nothing to do with peptide sequencing but rather describes a probabilistic process that transforms a peptide vector $P = (p_1, \ldots, p_m)$ into a *binary* vector $X = (x_1, \ldots, x_m)$ of the same length. We will see later how ideas developed for this model help us analyze real spectra.

We define the probability that P generates X as $\Pr(X|P) = \Pi_{i=1}^{m} \Pr(x_i|p_i)$, where $\Pr(x_i|p_i)$ is the probability that p_i in P generates x_i in X (Figure 11.22). For example, the probability that a 1 in P generates a 1 in X is written $\Pr(1|1)$ and is equal to some parameter ρ. The probability that a 0 in P generates a 1 in X is written $\Pr(1|0)$ and is equal to some parameter θ. The probability of a 1 in P generating a 0 in X is $\Pr(0|1) = 1 - \rho$, and the probability of a 0 in P generating a 0 in X is $\Pr(0|0) = 1 - \theta$.

For the toy amino acid alphabet containing just two amino acids with masses 2 and 3, Figure 11.23 illustrates the peptide vector $P = (0, 1, 0, 1, 0, 0, 1)$ generating the binary vector $X = (0, 0, 0, 1, 1, 0, 1)$ with probability

$$\Pr(X|P) = (1-\theta) \cdot (1-\rho) \cdot (1-\theta) \cdot \rho \cdot \theta \cdot (1-\theta) \cdot \rho.$$

WAS T. REX JUST A BIG CHICKEN?

$$
\begin{array}{c|cc}
 & \multicolumn{2}{c}{\text{symbol in } P} \\
 & 0 & 1 \\
\hline
\text{symbol in } X \quad 0 & 1-\vartheta & 1-\rho \\
1 & \vartheta & \rho
\end{array}
$$

FIGURE 11.22 A matrix describing a probabilistic process that transforms a peptide vector P into a binary vector X.

We are interested in the following problem.

Most Likely Peptide Vector Problem:

Find a most likely peptide vector for a given binary vector.

Input: A binary vector X and parameters ρ and θ such that $0 \leq \rho, \theta \leq 1$.
Output: A peptide vector P that maximizes $\Pr(X \mid P)$ as defined by probabilities ρ and θ among all possible peptide vectors.

The solution of this problem is given in **DETOUR: Solving the Most Likely Peptide Vector Problem**.

peptide vector P	0	1	0	1	0	0	1
binary vector X	0	0	0	1	1	0	1
$\Pr(X\|P)$	$\Pr(0\|0)$ $(1-\vartheta)$	$\cdot \Pr(0\|1)$ $(1-\rho)$	$\cdot \Pr(0\|0)$ $(1-\vartheta)$	$\cdot \Pr(1\|1)$ ρ	$\cdot \Pr(1\|0)$ ϑ	$\cdot \Pr(0\|0)$ $(1-\vartheta)$	$\cdot \Pr(1\|1)$ ρ
$\Pr(X\|\vec{0})$	$\Pr(0\|0)$ $(1-\vartheta)$	$\cdot \Pr(0\|0)$ $(1-\vartheta)$	$\cdot \Pr(0\|0)$ $(1-\vartheta)$	$\cdot \Pr(1\|0)$ ϑ	$\cdot \Pr(1\|0)$ ϑ	$\cdot \Pr(0\|0)$ $(1-\vartheta)$	$\cdot \Pr(1\|0)$ ϑ
LIKELIHOOD$(X\|P)$	1	$\cdot \frac{\Pr(0\|1)}{\Pr(0\|0)}$ $\frac{1-\rho}{1-\vartheta}$	$\cdot 1$	$\cdot \frac{\Pr(1\|1)}{\Pr(1\|0)}$ $\frac{\rho}{\vartheta}$	$\cdot 1$	$\cdot 1$	$\cdot \frac{\Pr(1\|1)}{\Pr(1\|0)}$ $\frac{\rho}{\vartheta}$
$\log_2($LIKELIHOOD$(X\|P))$	0	$+ \log_2 \frac{1-\rho}{1-\vartheta}$	$+ 0$	$+ \log_2 \frac{\rho}{\vartheta}$	$+ 0$	$+ 0$	$+ \log_2 \frac{\rho}{\vartheta}$

FIGURE 11.23 A peptide vector $P = (0, 1, 0, 1, 0, 0, 1)$ generates a binary vector $X = (0, 0, 0, 1, 1, 0, 1)$ with probability $\Pr(X|P)$.

The infinite monkey theorem

In Jonathan Swift's *Gulliver's Travels*, a professor of the Grand Academy of Legado asks his students to generate random strings of letters by turning cranks on a machine.

According to the professor, the Academy will eventually crank out brilliant works on all subjects.

Inspired by Swift's satirical treatment of certain academics, the **infinite monkey theorem** states that an immortal monkey typing an infinite sequence of symbols on a typewriter will one day reproduce *Hamlet*. In more technical terms, this theorem states that an infinite random string contains an arbitrary given text as a substring **almost surely**, or with probability equal to 1.

STOP and Think: In 2003, researchers placed a typewriter in a monkey enclosure, and found that the monkeys typed the letter "S" over and over. It is possible that an infinite random string generated by a monkey would contain only the letter "S". How, then, must this string contain *Hamlet* almost surely?

The probabilistic space of peptides in a spectral dictionary

In the main text, we defined the probability of *Peptide* as $1/20^{|Peptide|}$, and we defined the probability of a collection of peptides *Dictionary* as

$$\Pr(Dictionary) = \sum_{\text{each peptide } Peptide \text{ in } Dictionary} \frac{1}{20^{|Peptide|}}.$$

But why have we used the probabilistic notation? After all, consider the following exercise, which indicates that it is possible for $\Pr(Dictionary)$ to be larger than 1.

EXERCISE BREAK: If *Dictionary* is the set of all peptides of length at most 10, what is $\Pr(Dictionary)$?

Yet recall that a peptide can match a spectrum if and only if its mass is equal to the mass of the spectrum. Thus, no peptide in a *spectral* dictionary can contain another peptide in the dictionary as a substring, i.e., a spectral dictionary forms a **substring-free set**.

EXERCISE BREAK: Prove that if *Dictionary* is a substring-free set, then $\Pr(Dictionary) \leq 1$.

However, you may still be wondering which event corresponds to $\Pr(Dictionary)$. More precisely, what is the "probabilistic space" of underlying outcomes from which *Dictionary* is formed? The probabilistic space that we propose contains all decoy proteomes of length n, where n is the length of the longest peptide in a spectral dictionary *Dictionary*. In this space, we will assume that each decoy proteome has the same probability. Thus, our probabilistic space consists of 20^n elements, each with probability $1/20^n$. Note the change: instead of considering the probabilistic space of all *peptides* in *Dictionary*, we have switched to considering all *decoy proteomes*.

Each string *Peptide* in *Dictionary* appears in exactly $20^{n-|Peptide|}$ decoy proteomes as their first peptide. The combined probabilities of all these decoy proteomes sum to

$$20^{n-|Peptide|} \cdot \frac{1}{20^n} = \frac{1}{20^{|Peptide|}},$$

which is Pr(*Peptide*). Since spectral dictionaries are substring-free, each decoy proteome has at most one peptide from the spectral dictionary starting at its first position. Thus, Pr(*Dictionary*) is simply the combined probability of all decoy proteomes that begin with one of the peptides in *Dictionary*.

Are terrestrial dinosaurs really the ancestors of birds?

Aside from the mysterious presence of hemoglobin peptides in Asara's *T. rex* spectra, scientists have recently expressed doubts about the hypothesis that birds evolved from terrestrial dinosaurs like *T. rex* and that flight was achieved from the biophysically improbable **ground-up model**. This hypothesis assumes that, in order to evolve into birds, dinosaurs first must have reduced their size while simultaneously developing feathers (arguably the most complex evolutionary invention for flight).

Most early dinosaur studies favored evidence for a small arboreal animal as the more logical interpretation for the bird ancestor. This conjecture assumes that before evolving a system for sustained flight, early birds used gravity-assisted aerodynamics such as parachuting and gliding (the latter is used by modern flying squirrels). Thus, the tiny *Scansoriopteryx* (Figure 11.24), whose fossils contain impressions of feathers and whose foot adaptations indicate an arboreal lifestyle, competes with *T. rex* for the honor of being the ancestor of birds.

If you are interested in learning more about the evolutionary controversy surrounding the origin of birds, we suggest two papers, one on either side of the debate:

- "Jurassic archosaur is a non–dinosaurian bird" by Stephen Czerkas and Alan Feduccia.

- "Three crocodilian genomes reveal ancestral patterns of evolution among archosaurs" by Richard Green et al.

CHAPTER 11

FIGURE 11.24 An artistic recreation of *Scansoriopteryx*..

Solving the Most Likely Peptide Vector Problem

Define LIKELIHOOD$(X|P)$ as $\Pr(X|P)/\Pr(X|\vec{0})$, where $\vec{0}$ is an **all-zeroes vector** consisting of only zeroes. Figure 11.23 illustrates that a peptide vector $P = (0, 1, 0, 1, 0, 0, 1)$ generates a binary vector $X = (0, 0, 0, 1, 1, 0, 1)$ with

$$\text{LIKELIHOOD}(X|P) = \frac{\Pr(0|1)}{\Pr(0|0)} \cdot \frac{\Pr(1|1)}{\Pr(1|0)} \cdot \frac{\Pr(1|1)}{\Pr(1|0)}$$

$$= \frac{1-\rho}{1-\theta} \cdot \frac{\rho}{\theta} \cdot \frac{\rho}{\theta}.$$

To avoid dealing with extremely small values that result from many multiplications in LIKELIHOOD$(X | P)$, we will instead use the **log-likelihood** $\log_2(\text{LIKELIHOOD}(X|P))$. Finding a peptide vector maximizing the log-likelihood is equivalent to finding a most probable peptide vector. Figure 11.23 illustrates that the peptide vector $(0, 1, 0, 1, 0, 0, 1)$ generates a binary vector $(0, 0, 0, 1, 1, 0, 1)$ with log-likelihood equal to

$$\log_2 \frac{1-\rho}{1-\theta} + \log_2 \frac{\rho}{\theta} + \log_2 \frac{\rho}{\theta}.$$

We will now transform a binary vector $X = (x_1, \ldots, x_m)$ into a spectral vector $S = (s_1, \ldots, s_m)$ by changing each occurrence of 0 into the amplitude $\log_2[(1-\rho)/(1-\theta)]$ and each occurrence of 1 into the amplitude $\log_2(\rho/\theta)$. For example, the binary vector $(0, 0, 0, 1, 1, 0, 1)$ will be transformed into the spectral vector

$$\left(\log_2 \frac{1-\rho}{1-\theta}, \; \log_2 \frac{1-\rho}{1-\theta}, \; \log_2 \frac{1-\rho}{1-\theta}, \; \log_2 \frac{\rho}{\theta}, \; \log_2 \frac{\rho}{\theta}, \; \log_2 \frac{1-\rho}{1-\theta}, \; \log_2 \frac{\rho}{\theta} \right).$$

WAS T. REX JUST A BIG CHICKEN?

Note that $\log_2(\text{LIKELIHOOD}(X|P))$ is simply the dot product of the peptide vector $P = (p_1, \ldots, p_m)$ and the spectral vector $S = (s_1, \ldots, s_m)$,

$$P \cdot S = p_1 \cdot s_1 + \cdots + p_m \cdot s_m.$$

We denote $P \cdot S$ as $\text{SCORE}(P, S)$ (Figure 11.25) and define the score between a peptide vector and a spectral vector of different length as $-\infty$. We have therefore transformed the Most Likely Peptide Vector Problem into the Peptide Sequencing Problem.

peptide vector P	0	1	0	1	0	0	1
binary vector X	0	0	0	1	1	0	1
spectral vector S	$\log_2 \frac{1-\rho}{1-\theta}$	$\log_2 \frac{1-\rho}{1-\theta}$	$\log_2 \frac{1-\rho}{1-\theta}$	$\log_2 \frac{\rho}{\theta}$	$\log_2 \frac{\rho}{\theta}$	$\log_2 \frac{1-\rho}{1-\theta}$	$\log_2 \frac{\rho}{\theta}$
$\text{SCORE}(P, S)$		$\log_2 \frac{1-\rho}{1-\theta}$	+	$\log_2 \frac{\rho}{\theta}$	+		$\log_2 \frac{\rho}{\theta}$

FIGURE 11.25 Scoring a peptide vector P against a spectral vector S as the dot product $\text{SCORE}(P, S)$.

Our conversion of a binary vector X into a spectral vector S was based on a simple probabilistic model (describing how a peptide vector generates a binary vector) with just two parameters, ρ and θ. To see how spectra are transformed into spectral vectors in practice, see **DETOUR: Selecting Parameters for Transforming Spectra into Spectral Vectors**.

Selecting parameters for transforming spectra into spectral vectors

If mass spectrometers generated binary spectra, then we could start by forming a large **training sample** of annotated spectra for which the peptides that generated these spectra are known. We could then estimate ρ (as the frequency of ones in binary spectra being generated by ones in peptide vectors) and θ (as the frequency of ones in binary spectra being generated by zeroes in peptide vectors) across all annotated spectra in the training sample. But since real mass spectrometers generate *integer-valued* rather than *binary* spectra, deriving parameters becomes more complex.

STOP and Think: Can you devise a probabilistic model that would convert the peptide vector (0, 1, 0, 1, 0, 0, 1) into the integer-valued vector (3, 4, 2, 6, 9, 4, 7)?

However, a similar probabilistic model will work if we define the probability of converting zeroes and ones in the peptide vector into various intensities in the real spectra (rather than into zeroes and ones as before). In fact, the conversion of real spectra into spectral vectors is based on a similar log-likelihood model that uses dozens of probabilistic parameters. Algorithms for converting real spectra into spectral vectors attempt to optimize these parameters so that amplitudes at prefix coordinates are maximized and amplitudes at noise coordinates are minimized.

To derive these parameters, we again need to build a large training sample of annotated spectra. We can consider all peaks with a certain intensity level in all spectra and compute which fraction of them are annotated by prefix or suffix peptides. For example, only 19% and 45% of the ten highest intensity peaks in Collision-Induced Dissociation spectra (which are similar to ones generated in Asara's laboratory) are explained by prefix and suffix peptides, respectively. The remaining high-intensity peaks are treated as noise. Given a spectrum generated by an unknown peptide vector $P = (p_1, \ldots, p_m)$, its spectral vector $S = (s_1, \ldots, s_m)$ is derived using these frequencies so that s_i is the log likelihood ratio $\log_2(\text{Pr}_1/\text{Pr}_0)$, where Pr_1 is an estimate of the probability that $p_i = 1$, and Pr_0 is an estimate of the probability that $p_i = 0$. A complete discussion of the details of the algorithm for generating spectral vectors is beyond the scope of this detour.

Figure 11.26 shows the set of prefix masses for *DinosaurPeptide* along with amplitudes of the spectral vector corresponding to these masses. Note from Figure 11.8 (bottom) that most amplitudes of the spectral vector are negative. The blue elements in Figure 11.26 correspond to positions that significantly exceed the average amplitude value.

	1	2	3	4	5	6	7	8	9	10	11	12	13	14
amino acid	G	L	V	G	A	P	G	L	R	G	L	P	G	K
mass	57	113	99	57	71	97	57	113	156	57	113	97	57	128
prefix mass	57	170	269	326	397	494	551	664	820	877	990	1087	1144	1272
amplitude	-8	+1	-4	-6	-6	+3	+1	-4	-8	+18	+11	-10	-7	0

FIGURE 11.26 Masses of amino acids in GLVGAPGLRGLPGK (second line), prefix masses for this peptide (third line), and the corresponding elements of the spectral vector for *DinosaurSpectrum*, shown in Figure 11.5 (top) (fourth line). Blue elements correspond to amplitudes that are significantly higher than the average, which is negative. Note that the second and third tallest peaks in the spectrum (labeled b_{10} and b_{11}) correspond to maximum amplitudes $+18$ and $+11$ in the spectral vector of *DinosaurSpectrum*. Also note that because there is no peak b_{12} (or y_2) in Figure 11.5 (bottom), $s_{1087} = -10$ is very small.

Bibliography Notes

T. rex peptides were reported by Asara et al., 2007 and faced criticism in Pevzner, Kim, and Ng, 2008, Buckley et al., 2008, and Kim et al., 2015. The "dinosaur DNA" paper by Woodward, Weyand, and Bunnell, 1994 was refuted by Hedges and Schweitzer, 1995. Czerkas and Feduccia, 2014 recently argued that birds did not evolve from dinosaurs, while Green et al., 2014 recently argued otherwise.

Chen et al., 2001 solved the Anti-Symmetric Path Problem in the case of graphs arising from mass spectra. Searching a protein database for the purpose of peptide identification in mass spectrometry was pioneered by Eng, McCormack, and Yates, 1994. The spectral alignment algorithm was introduced by Pevzner, Dančík, and Tang, 2000. The concept of spectral dictionary and the algorithm for evaluating statistical significance of PSMs were introduced by Kim, Gupta, and Pevzner, 2008 and Kim et al., 2009.

Bibliography

Aho, A. V. and M. J. Corasick (1975). "Efficient String Matching: An Aid to Bibliographic Search". *Communications of the ACM* Vol. 18: 333–340.

Alon, U., N. Barkai, D. A. Notterman, K. Gish, S. Ybarra, D. Mack, and A. J. Levine (1999). "Broad patterns of gene expression revealed by clustering analysis of tumor and normal colon tissues probed by oligonucleotide arrays". *Proceedings of the National Academy of Sciences* Vol. 96: 6745–6750.

Altschul, S. F., W. Gish, W. Miller, E. W. Myers, and D. J. Lipman (1990). "Basic local alignment search tool." *Journal of Molecular Biology* Vol. 215: 403–410.

Arthur, D. and S. Vassilvitskii (2007). "k-means++: The Advantages of Careful Seeding". In: *Proceedings of the Eighteenth Annual ACM-SIAM Symposium on Discrete Algorithms*. New Orleans, Louisiana: Society for Industrial and Applied Mathematics, 1027–1035.

Asara, J. M., M. H. Schweitzer, L. M. Freimark, M. Phillips, and L. C. Cantley (2007). "Protein Sequences from Mastodon and Tyrannosaurus rex Revealed by Mass Spectrometry". *Science* Vol. 316: 280–285.

Baldi, P., Y. Chauvin, T. Hunkapiller, and M. A. McClure (1994). "Hidden Markov models of biological primary sequence information". *Proceedings of the National Academy of Sciences* Vol. 91: 1059–1063.

Bateman, A., E. Birney, L. Cerruti, R. Durbin, L. Etwiller, S. R. Eddy, S. Griffiths-Jones, K. L. Howe, M. Marshall, and E. L. L. Sonnhammer (2002). "The Pfam Protein Families Database". *Nucleic Acids Research* Vol. 30: 276–280.

Baum, L. E., T. Petrie, G. Soules, and N. Weiss (1970). "A Maximization Technique Occurring in the Statistical Analysis of Probabilistic Functions of Markov Chains". *The Annals of Mathematical Statistics* Vol. 41: pp. 164–171.

Beerenwinkel, N. and M. Drton (2007). "A mutagenetic tree hidden Markov model for longitudinal clonal HIV sequence data". *Biostatistics* Vol. 8: 53–71.

Ben-Dor, A., R. Shamir, and Z. Yakhini (1999). "Clustering Gene Expression Patterns". *Journal of Computational Biology* Vol. 6: 281–297.

Bezdek, J. C. (1981). *Pattern Recognition with Fuzzy Objective Function Algorithms*. Kluwer Academic Publishers.

Buckley, M., A. Walker, S. Y. W. Ho, Y. Yang, C. Smith, P. Ashton, J. T. Oates, E. Cappellini, H. Koon, K. Penkman, B. Elsworth, D. Ashford, C. Solazzo, P. Andrews, J. Strahler, B. Shapiro, P. Ostrom, H. Gandhi, W. Miller, B. Raney, M. I. Zylber, M. T. P. Gilbert, R. V. Prigodich, M. Ryan, K. F. Rijsdijk, A. Janoo, and M. J. Collins (2008). "Comment on 'Protein Sequences from Mastodon and Tyrannosaurus rex Revealed by Mass Spectrometry'". *Science* Vol. 319: 33.

Burrows, D. and M. J. Wheeler (1994). "A block sorting lossless data compression algorithm". *Technical Report 124, Digital Equipment Corporation*.

Cann, R. L., M. Stoneking, and A. C. Wilson (1987). "Mitochondrial DNA and human evolution." *Nature* Vol. 325: 31–36.

Ceppellini, B. R., M. Siniscalco, and C. A. B. Smith (1955). "The Estimation of Gene Frequencies in a Random-Mating Population". *Annals of Human Genetics* Vol. 20: 97–115.

Chen, T., M.-Y. Kao, M. Tepel, J. Rush, and G. M. Church (2001). "A Dynamic Programming Approach to De Novo Peptide Sequencing via Tandem Mass Spectrometry". *Journal of Computational Biology* Vol. 8: 325–337.

Churchill, G. (1989). "Stochastic models for heterogeneous DNA sequences". *Bulletin of Mathematical Biology* Vol. 51: 79–94.

Clayton-Smith, J., J. O'Sullivan, S. Daly, S. Bhaskar, R. Day, B. Anderson, A. K. Voss, T. Thomas, L. G. Biesecker, P. Smith, A. Fryer, K. E. Chandler, B. Kerr, M. Tassabehji, S. A. Lynch, M. Krajewska-Walasek, S. McKee, J. Smith, E. Sweeney, S. Mansour, S. Mohammed, D. Donnai, and G. Black (2011). "Whole-exome-sequencing identifies mutations in histone acetyltransferase gene KAT6B in individuals with the Say-

Barber-Biesecker variant of Ohdo syndrome". *The American Journal of Human Genetics* Vol. 89: 675–681.

Cristianini, N. and M. W. Hahn (2007). *Introduction to Computational Genomics*. Cambridge University Press.

Czerkas, S. and A. Feduccia (2014). "Jurassic archosaur is a non-dinosaurian bird". *Journal of Ornithology* Vol. 155: 841–851.

De Jong, J., A. De Ronde, W. Keulen, M. Tersmette, and J. Goudsmit (1992). "Minimal Requirements for the Human Immunodeficiency Virus Type 1 V3 Domain To Support the Syncytium-Inducing Phenotype: Analysis by Single Amino Acid Substitution". *Journal of Virology* Vol. 66: 6777–6780.

DeRisi, J. L., V. R. Iyer, and P. O. Brown (1997). "Exploring the metabolic and genetic control of gene expression on a genomic scale". *Science* Vol. 278: 680–686.

Do, C. B. and S. Batzoglou (2008). "What is the expectation maximization algorithm?" *Nature Biotechnology* Vol. 26: 897–899.

Eisen, M. B., P. T. Spellman, P. O. Brown, and D. Botstein (1998). "Cluster analysis and display of genome-wide expression patterns". *Proceedings of the National Academy of Sciences* Vol. 95: 14863–14868.

Eng, J. K., A. L. McCormack, and J. R. Yates (1994). "An approach to correlate tandem mass spectral data of peptides with amino acid sequences in a protein database". *Journal of the American Society for Mass Spectrometry* Vol. 5: 976–989.

Felsenstein, J. (2004). *Inferring Phylogenies*. Sinauer Associates.

Ferragina, P. and G. Manzini (2000). "Opportunistic Data Structures with Applications". In: *Proceedings of the 41st Annual Symposium on Foundations of Computer Science*. IEEE Computer Society, 390–398.

Gao, F., E. Bailes, D. L. Robertson, Y. Chen, C. M. Rodenburg, S. F. Michael, L. B. Cummins, L. O. Arthur, M. Peeters, G. M. Shaw, P. M. Sharp, and B. H. Hahn (1999). "Origin of HIV-1 in the chimpanzee Pan troglodytes troglodytes". *Nature* Vol. 397: 436–441.

Green, R. E., E. L. Braun, J. Armstrong, D. Earl, N. Nguyen, G. Hickey, M. W. Vandewege, J. A. St. John, S. Capella-Gutiérrez, T. A. Castoe, C. Kern, M. K. Fujita, J. C. Opazo, J. Jurka, K. K. Kojima, J. Caballero, R. M. Hubley, A. F. Smit, R. N. Platt, C. A. Lavoie, M. P. Ramakodi, J. W. Finger, A. Suh, S. R. Isberg, L. Miles, A. Y. Chong,

W. Jaratlerdsiri, J. Gongora, C. Moran, A. Iriarte, J. McCormack, S. C. Burgess, S. V. Edwards, E. Lyons, C. Williams, M. Breen, J. T. Howard, C. R. Gresham, D. G. Peterson, J. Schmitz, D. D. Pollock, D. Haussler, E. W. Triplett, G. Zhang, N. Irie, E. D. Jarvis, C. A. Brochu, C. J. Schmidt, F. M. McCarthy, B. C. Faircloth, F. G. Hoffmann, T. C. Glenn, T. Gabaldón, B. Paten, and D. A. Ray (2014). "Three crocodilian genomes reveal ancestral patterns of evolution among archosaurs". *Science* Vol. 346:

Hedges, S. and M. Schweitzer (1995). "Detecting dinosaur DNA". *Science* Vol. 268: 1191–1192.

Kellis, M., B. W. Birren, and E. S. Lander (2004). "Proof and evolutionary analysis of ancient genome duplication in the yeast Saccharomyces cerevisiae". *Nature* Vol. 428: 617–624.

Kim, S., M. J. Fitzgibbon, M. W. McIntosh, and P. A. Pevzner (2015). "Tyrannosaurus rex protein sequencing controversy: contamination or treasure trove of ancient proteins?" *(unpublished manuscript)*.

Kim, S., N. Gupta, and P. A. Pevzner (2008). "Spectral Probabilities and Generating Functions of Tandem Mass Spectra: A Strike against Decoy Databases". *Journal of Proteome Research* Vol. 7: PMID: 18597511, 3354–3363.

Kim, S., N. Gupta, N. Bandeira, and P. A. Pevzner (2009). "Spectral dictionaries: Integrating de novo peptide sequencing with database search of tandem mass spectra." *Molecular & Cellular Proteomics* Vol. 8: 53–69.

Krogh, A., M. Brown, I. Saira Mian, K. Sjolander, and D. Haussler (1994). "Hidden Markov Models in Computational Biology: Applications to Protein Modeling". *Journal of Molecular Biology* Vol. 235: 1501–1531.

Lloyd, S. (1982). "Least squares quantization in PCM". *IEEE Transactions on Information Theory* Vol. 28: 129–137.

Manber, U. and G. Myers (1990). "Suffix Arrays: A New Method for On-line String Searches". In: *Proceedings of the First Annual ACM-SIAM Symposium on Discrete Algorithms*. Society for Industrial and Applied Mathematics, 319–327.

Martin, N., E. Ruedi, R. LeDuc, F.-J. Sun, and G. Caetano-Anolles (2007). "Gene-interleaving patterns of synteny in the Saccharomyces cerevisiae genome: Are they proof of an ancient genome duplication event?" *Biology Direct* Vol. 2: 23.

Metzker, M. L., D. P. Mindell, X.-M. Liu, R. G. Ptak, R. A. Gibbs, and D. M. Hillis (2002). "Molecular evidence of HIV-1 transmission in a criminal case". *Proceedings of the National Academy of Sciences* Vol. 99: 14292–14297.

Montgomery, S. B., D. Goode, E. Kvikstad, C. A. Albers, Z. Zhang, X. J. Mu, G. Ananda, B. Howie, K. J. Karczewski, K. S. Smith, V. Anaya, R. Richardson, J. Davis, D. G. MacArthur, A. Sidow, L. Duret, M. Gerstein, K. Markova, J. Marchini, G. A. McVean, and G. Lunter (2013). "The origin, evolution and functional impact of short insertion-deletion variants identified in 179 human genomes". *Genome Research* Vol. 23: 749–761.

O'Brien, S. J., W. G. Nash, D. E. Wildt, M. E. Bush, and R. E. Benveniste (1985). "A molecular solution to the riddle of the giant panda's phylogeny". *Nature* Vol. 317: 140–144.

Ohno, S. (1970). *Evolution by Gene Duplication*. Springer-Verlag.

Pevzner, P. A., V. Dančík, and C. L. Tang (2000). "Mutation-Tolerant Protein Identification by Mass Spectrometry". *Journal of Computational Biology* Vol. 7: 777–787.

Pevzner, P. A., S. Kim, and J. Ng (2008). "Comment on 'Protein sequences from mastodon and Tyrannosaurus rex revealed by mass spectrometry'". *Science* Vol. 321: 1040.

Robinson, D. F. (1971). "Comparison of labeled trees with valency three". *Journal of Combinatorial Theory, Series B* Vol. 11: 105–119.

Saitou, N. and M. Nei (1987). "The neighbor-joining method: a new method for reconstructing phylogenetic trees." *Mol Biol Evol* Vol. 4: 406–425.

Sankoff, D. (1975). "Minimal Mutation Trees of Sequences". *SIAM Journal on Applied Mathematics* Vol. 28: 35–42.

Sokal, R. R. and C. D. Michener (1958). "A statistical method for evaluating systematic relationships". *University of Kansas Scientific Bulletin* Vol. 28: 1409–1438.

Studier, J. A. and K. J. Keppler (1988). "A note on the neighbor-joining algorithm of Saitou and Nei." *Mol Biol Evol* Vol. 5: 729–731.

Thomson, J. M., E. A. Gaucher, M. F. Burgan, D. W. De Kee, T. Li, J. P. Aris, and S. A. Benner (2005). "Resurrecting ancestral alcohol dehydrogenases from yeast." *Nature Genetics* Vol. 37: 630–635.

Tuzun, E., A. J. Sharp, J. A. Bailey, R. Kaul, V. A. Morrison, L. M. Pertz, E. Haugen, H. Hayden, D. Albertson, D. Pinkel, M. V. Olson, and E. E. Eichler (2005). "Fine-scale structural variation of the human genome". *Nature Genetics* Vol. 37: 727–732.

Viterbi, A. (1967). "Error bounds for convolutional codes and an asymptotically optimum decoding algorithm". *IEEE Transactions on Information Theory* Vol. 13: 260–269.

Weiner, P. (1973). "Linear Pattern Matching Algorithms". In: *Proceedings of the 14th Annual Symposium on Switching and Automata Theory*. IEEE Computer Society, 1–11.

Whiting, M. F., S. Bradler, and T. Maxwell (2003). "Loss and recovery of wings in stick insects". *Nature* Vol. 421: 264–267.

Wolfe, K. H. and D. C. Shields (1997). "Molecular evidence for an ancient duplication of the entire yeast genome". *Nature* Vol. 387: 708–713.

Woodward, N. Weyand, and M Bunnell (1994). "DNA sequence from Cretaceous period bone fragments". *Science* Vol. 266: 1229–1232.

Zaretskii, Z. A. (1965). "Constructing a tree on the basis of a set of distances between the hanging vertices". *Uspekhi Mat. Nauk* Vol. 20: 90–92.

Zuckerkandl, E. and L. Pauling (1965). "Molecules as documents of evolutionary history." *Journal of Theoretical Biology* Vol. 8: 357–366.

Image Courtesies

Figure 7.1 (right): Luc Viatour
Figure 7.14: Wikimedia Commons user Praveenp
Figure 7.15: Julie Langford (squirrel monkey), André Karwath (old world monkey), Frans de Waal (chimp), Kabir Bakie (gorilla)
Figure 7.21: Wikimedia Commons user Drägüs (winged stick insect), L. Shyamal (wingless stick insect), Katka Nemčoková (giant centipede)
Figure 7.30: Frans de Waal (chimp), Kabir Bakie (gorilla)
Figure 7.36: Wikimedia Commons user Praveenp
Figure 7.41: Alan D. Wilson (polar bear), Mark Dumont (spectacled bear), Peter Meenen (red panda), Wikimedia commons user Darkone (raccoon).
Figure 11.24: Matt Martyniuk
Figure 8.27: Guillaume Paumier (microarray)
Figure 10.27: Thomas Splettstoesser
Figure 10.28 (top): Wikimedia Commons user Dna 621